# The Galactic Center
(California Institute of Technology, 1982)

**AIP Conference Proceedings**
Series Editor: Hugh C. Wolfe
**Number 83**

# The Galactic Center
(California Institute of Technology, 1982)

Editors
**Guenter R. Riegler**
Jet Propulsion Laboratory, CIT
and
**Roger D. Blandford**
California Institute of Technology

**American Institute of Physics**
New York 1982

Copying fees: The code at the bottom of the first page of each article in this volume gives the fee for each copy of the article made beyond the free copying permitted under the 1978 US Copyright Law. (See also the statement following "Copyright" below). This fee can be paid to the American Institute of Physics through the Copyright Clearance Center, Inc., Box 765, Schenectady, N.Y. 12301.

Copyright © 1982 American Institute of Physics

Individual readers of this volume and non-profit libraries, acting for them, are permitted to make fair use of the material in it, such as copying an article for use in teaching or research. Permission is granted to quote from this volume in scientific work with the customary acknowledgment of the source. To reprint a figure, table or other excerpt requires the consent of one of the original authors and notification to AIP. Republication or systematic or multiple reproduction of any material in this volume is permitted only under license from AIP. Address inquiries to Series Editor, AIP Conference Proceedings, AIP, 335 E. 45th St., New York, N. Y. 10017

L.C. Catalog Card No. 82-071635
ISBN 0-88318-182-7
DOE CONF- 820139

PREFACE

The Workshop on the Galactic Center was held at the California Institute of Technology on January 7 and 8, 1982. Its purpose was to review observations and models of the Galactic Center, and to encourage an exchange of opinions among persons interested in physical processes and observable features of the Galactic Center. The concept for this workshop was formed when new discoveries prompted researchers in gamma-ray astronomy to contact their colleagues for the latest results in radio and infrared astronomy, and vice versa.

In this volume, the authors of both the invited and contributed papers were given the opportunity to elaborate upon their talks to provide more detail than could be presented at the meeting. We hope that the publication of these proceedings will serve as stimulus to further research.

The workshop organizing committee consisted of Roger D. Blandford, California Institute of Technology; John Lacy, California Institute of Technology; Richard E. Lingenfelter, University of California at San Diego; K. Y. Lo, California Institute of Technology; Gerry Neugebauer, California Institute of Technology; and Guenter R. Riegler, Jet Propulsion Laboratory (Chairman).

We are particularly grateful to Margaret Katz, who coordinated the many arrangements for the workshop. Support for the Workshop and the proceedings was provided by the National Aeronautics and Space Administration and the California Institute of Technology President's Fund.

## Participants in the Workshop

| | |
|---|---|
| David Aitken | - UCL, London |
| Philip Anderson | - Caltech |
| Gordon G. Augason | - NASA/ARC |
| Don Backer | - UC Berkeley |
| Taeil Bai | - UC San Diego |
| Eric Becklin | - U of Hawaii |
| Roger Blandford | - Caltech |
| Robert L. Brown | - NRAO |
| Michael L. Burns | - NASA/GSFC |
| Marshall Cohen | - Caltech |
| Walter Cook | - Caltech |
| R. D. Davies | - U of Manchester, U.K. |
| Bruce Dayton | - UC Riverside |
| William Dent | - U of Massachusetts |
| Harriet Dinerstein | - NASA/ARC |
| Luke Dones | - UC Berkeley |
| Alfred Dunklee | - Jet Propulsion Laboratory |
| Edward Edelson | - Mosaic Magazine |
| Dan Forrest | - U of New Hampshire |
| Yasuo Fukui | - Nagoya University |
| Barbara Gardner | - U of New Hampshire |
| Ian Gatley | - UKIRT, Hawaii |
| Tom Geballe | - UKIRT, Hawaii |
| Barry Geldzahler | - NRL |
| Reinhard Genzel | - UC Berkeley |
| Duane Gerber | - UC San Diego |
| Rolf Güsten | - Max-Planck Institut, Bonn |
| Paul Harvey | - U of Texas |
| Hugh Hudson | - UC San Diego |
| Allan S. Jacobson | - Jet Propulsion Laboratory |
| Dayton Jones | - Caltech |
| Jocelyn Keene | - Caltech |
| Donald A. Kniffen | - GSFC |
| Tom Kuiper | - Jet Propulsion Laboratory |
| John Lacy | - Caltech |
| George Lake | - Bell Labs |
| Jess L. Lang | - UC Los Angeles |
| Marcia Lebofsky | - Steward Obs., U of Az |
| Marvin Leventhal | - Bell Labs |
| Robert Lin | - UC Berkeley |
| James C. Ling | - Jet Propulsion Laboratory |
| Richard Lingenfelter | - UC San Diego |
| Harvey Liszt | - NRAO |
| William A. Mahoney | - Jet Propulsion Laboratory |
| Alan P. Marscher | - Boston U |
| Toshio Matsumoto | - Nagoya University |
| Jim Matteson | - UC San Diego |
| Peter McGregor | - Mt. Wilson |
| Albert Metzger | - Jet Propulsion Laboratory |
| Alan I. Moffett | - Caltech |
| Gerry Neugebauer | - Caltech |
| Sten Odenwald | - Harvard U |
| Jan H. Oort | - Leiden Observatory |

| | |
|---|---|
| W. S. Paciesas | - NASA/GSFC |
| J. Pier | - Caltech |
| Tom Prince | - Caltech |
| Raymond J. Proctor | - UC San Diego |
| Reuven Ramaty | - NASA/GSFC |
| A. Readhead | - Caltech |
| Martin Rees | - Cambridge University |
| Stephen P. Reynolds | - U of Virginia |
| George R. Ricker | - MIT |
| Guenter R. Riegler | - Jet Propulsion Laboratory |
| George Rieke | - Steward Obs., U of Az. |
| Jim Roberts | - Pallas Athene Films |
| Ray Russell | - Aerospace Corporation |
| O. Rydbeck | - Onsala Space Observatorium |
| Aage Sandqvist | - Stockholm Observatory |
| Howard Smith | - NRL |
| Philip Solomon | - SUNY Stony Brook |
| Mark Stier | - NASA/GSFC |
| John Storey | - Anglo Australian Observatory |
| Ed Sutter | - Caltech |
| Bonnard Teegarden | - NASA/GSFC |
| Kip S. Thorne | - Caltech |
| Charles Townes | - UC Berkeley |
| Virginia Trimble | - Caltech |
| Jack Tueller | - NASA/GSFC |
| T. Tumer | - UC Riverside |
| Steve Univin | - Caltech |
| J. M. van der Hulst | - U of Minnesota |
| Larry Varnell | - Jet Propulsion Laboratory |
| Dan Watson | - UC Berkeley |
| V. Weideman | - Caltech |
| W. A. Wheaton | - Jet Propulsion Laboratory |
| Bill White | - UC Los Angeles |
| Jim Willett | - Jet Propulsion Laboratory |
| Steven Willner | - CFA |
| Al Wootten | - Caltech |
| Diana Worrall | - UC San Diego |
| Ed Zanrosso | - UC Riverside |
| Allen Zych | - UC Riverside |

TABLE OF CONTENTS

## I. RADIO OBSERVATIONS

OBSERVATIONS OF THE GALACTIC CENTER IN THE RADIO BAND
K. Y. Lo .................................................... 1

MOLECULAR GAS NEAR THE GALACTIC CENTER - RECENT
CONTRIBUTIONS FROM RADIOASTRONOMY*
R. Güsten ................................................... 9

THE RELATIVE LOCATIONS OF THE SGR A MOLECULAR CLOUDS AND
CONTINUUM SOURCES*
Aa. Sandqvist ............................................... 12

DEPRESSION OF MOLECULAR EMISSION IN THE LINE OF SIGHT OF
SGR A WEST*
Y. Fukui, H. Ogawa, S. Deguchi and H. Suzuki ................ 18

## II. INFRARED CONTINUUM OBSERVATIONS

INFRARED OBSERVATIONS OF THE GALACTIC CENTER
I. Gatley ................................................... 25

NEW FAR INFRARED OBSERVATIONS OF THE CENTRAL 30' OF THE GALAXY*
W. A. Dent, M. W. Werner, I. Gatley, E. E. Becklin,
R. H. Hilderbrand, J. Keene and S. E. Whitcomb .............. 33

LARGE BEAM OBSERVATIONS OF THE GALACTIC CENTER AT 150,
200 AND 300 $\mu$m*
M. T. Stier, E. Dwek, R. F. Silverberg, M. G. Hauser,
L. Cheung, T. Kelsall and D. Y. Gezari ...................... 42

BALOON OBSERVATION OF THE CENTRAL BULGE OF OUR GALAXY
IN NEAR INFRARED RADIATION*
T. Matsumoto, S. Hayakawa, H. Koizumi, H. Murakami,
K. Uyama, T. Yamagami and J. A. Thomas ...................... 48

---

\* Contributed paper

III. INFRARED SPECTROSCOPY

INFRARED OBSERVATIONS OF THE IONIZED GAS IN THE
GALACTIC CENTER
    J. H. Lacy .................................................. 53

B$\alpha$ AND Ne II LINE SPECTROSCOPY IN THE VICINITY OF THE GALACTIC
CENTER SOURCE IRS 16*
    T. R. Geballe, S. E. Persson, J. H. Lacy,
    G. Neugebauer and S. C. Beck ............................... 60

SPATIAL AND SPECTRAL STUDIES OF THE GALACTIC CENTER
NEAR 10$\mu$*
    D. K. Aitken, M. C. Allen and P. F. Roche .................. 67

OI AND OIII IN SGR A: NEUTRAL AND IONIZED GAS AT THE
GALACTIC CENTER*
    R. Genzel, D. Watson, C. Townes, D. Lester,
    H. Dinerstein, M. Werner and J. Storey ..................... 72

THREE COMPACT SOURCES WITH UNUSUAL 2 TO 4 MICRON SPECTRA*
    S. P. Willner and J. L. Pipher ............................. 77

TWO MICRON OBSERVATIONS OF $^{12}$CO AND $^{13}$CO IN THE RED GIANT
SOURCES IRS 7, IRS 12, and IRS 19*
    G. C. Augason, H. A. Smith, E. R. Wollman,
    H. P. Larson and H. R. Johnson ............................. 82

IV. INFRARED IMAGING OBSERVATIONS

MAPPING AND IMAGING OF THE GALACTIC CENTER IN THE NEAR INFRARED*
    J. W. V. Storey ............................................ 85

TWO COLOR CCD OBSERVATIONS OF THE GALACTIC CENTER*
    J. A. Biretta, K. Y. Lo and P. J. Young .................... 91

DISCOVERY OF THREE NEAR INFRARED OBJECTS IN CCD IMAGES OF
THE GALACTIC CENTER*
    G. R. Ricker, M. W. Bautz, D. L. DePoy and S. S. Meyer ....... 97

THE POSITION OF THE INFRARED SOURCE IRS 16 IN THE GALACTIC
CENTER REGION RELATIVE TO A VISUAL FIELD STAR*
    G. Neugebauer, K. Matthews and B. T. Soifer ................ 107

---

* Contributed paper

V. GAMMA-RAY OBSERVATIONS

OBSERVATIONS OF CONTINUUM X-RAY AND GAMMA-RAY EMISSION
FROM THE GALACTIC CENTER
    J. L. Matteson ................................................. 109

OBSERVATIONS OF GAMMA-RAY LINE EMISSION FROM THE
GALACTIC CENTER REGION
    A. S. Jacobson ................................................. 123

TIME VARIABLE POSITRON ANNIHILATION RADIATION FROM THE
GALACTIC CENTER DIRECTION*
    M. Leventhal and C. J. MacCallum .............................. 132

OBSERVATIONS OF THE GALACTIC CENTER WITH THE GSFC LOW-ENERGY
GAMMA-RAY SPECTROMETER: PRELIMINARY RESULTS*
    W. S. Paciesas, T. L. Cline, B. J. Teegarden,
    J. Tueller, P. Durouchoux and J. M. Hameury .................. 139

EMISSION IN THE 0.3 TO 1.0 MeV RANGE FROM THE GALACTIC
CENTER REGION*
    B. M. Gardner, D. J. Forrest, P. P. Dunphy and E. L. Chupp ...144

VI. THEORY: MODELS FOR THE COMPACT SOURCE

ON THE ORIGIN OF THE POSITRON ANNIHILATION RADIATION FROM
THE DIRECTION OF THE GALACTIC CENTER
    R. E. Lingenfelter and R. Ramaty ............................. 148

THE INTENSITY AND SPECTRUM OF GALACTIC CENTER $\beta^+$
ANNIHILATION PHOTONS AFTER COMPTON SCATTERING*
    D. J. Forrest ................................................. 160

PHYSICS OF BLACK HOLES
    K. S. Thorne .................................................. 165

THE COMPACT SOURCE AT THE GALACTIC CENTER
    M. J. Rees .................................................... 166

POSITRON PRODUCTION NEAR A $10^6$ $M_\odot$ BLACK HOLE*
    R. D. Blandford ............................................... 177

---

* Contributed paper

VII. THEORY: DYNAMICS OF THE GALACTIC CENTER

    GAS MOTIONS IN THE CENTRAL REGION AND THEIR INTERPRETATION
        J. H. Oort .................................................. 180

    TRIAXIALITY AND THE GALACTIC CENTER*
        G. R. Lake and C. Norman ..................................... 189

VIII. COMPARISON WITH OTHER GALAXIES

    COMPARISON OF THE GALACTIC CENTER WITH OTHER GALAXIES
        G. H. Rieke and M. J. Lebofsky ............................... 194

IX. FUTURE DIRECTIONS

    RADIO OBSERVATIONS OF THE GALACTIC CENTER: FUTURE DIRECTIONS
        R. L. Brown .................................................. 204

    FUTURE DIRECTIONS IN X-RAY/GAMMA-RAY OBSERVATIONS
        D. A. Kniffen ................................................ 208

---

\* Contributed paper

Chapter I        Radio Observations

## OBSERVATIONS OF THE GALACTIC CENTER IN THE RADIO BAND

K. Y. Lo

Department of Astronomy
California Institute of Technology

### ABSTRACT

Observations of radio frequency spectral line of atomic hydrogen, ionized hydrogen, and molecular gas have long indicated the presence of large radial motions and large velocity dispersion of the gas in the central few kiloparsecs. The large non-circular motion of a large portion of the gas has been interpreted as evidence for expulsion from the center, and the large velocity dispersion within the central parsec as possible evidence for a massive collapsed object.

The presence in the center of a compact nonthermal radio source that may be a very weak version of extragalactic compact radio sources found in quasars and radio galaxies further suggests unusual activities in the actual center.

We review here the observations and properties of the compact radio source at the galactic center and the distribution of the ionized gas within a parsec of the source. We shall also discuss the constraints on the nature of the underlying energy source.

### INTRODUCTION

The highly obscured center of our Galaxy is particularly suited to radio wavelength observations which are unaffected by intervening dust. Soon after the discovery from interstellar space of the $\lambda=21$ cm ground-state hyperfine structure transition of neutral hydrogen (HI), observations of the galactic center region were made and large scale expanding motion of the gas was discovered[1]. More recent observations of transitions in molecules such as $H_2CO$[2], $NH_3$[3], and CO [4] have indicated similar expanding motions in the molecular gas. As the gas motions in the central few kiloparsecs and their interpretation will be discussed in greater detail by Professor Oort later in the proceeding [5], suffice it to say here that the expanding motions of the gas have been cited as evidence for expulsion of matter from the center (cf. Oort[6]).

Observations by Pauls et al.[7] of a very large line-width ($\Delta v \simeq 200$ Km/s) in the radio recombination line, H109$\alpha$, from Sgr A West,[8] the thermal component of the radio source at the center, provided the first indication that the ionized gas within the central few parsecs must have large non-random motions. Subsequent higher spatial resolution studies of the ionized gas within the central parsec, derived from observations of the 12.8 $\mu$ NeII fine structure line[9], will be discussed by Lacy[10]. If the ionized gas motions were governed by gravitation, the large velocity dispersion may imply the presence of a collapsed mass of $<5 \times 10^6$ $M_\odot$.

In the meantime, high angular resolution observations of the radio continuum from Sgr A, the radio source at the galactic center, have revealed the presence of a compact nonthermal radio source that may be identified as the center of activity.

We review in the next section the observations leading to the identification of the three major components of Sgr A. In the following section, we shall review the observations and properties of the compact nonthermal radio source.

## SAGITTARIUS A

Figure 1 shows an intensity map of the radio continuum emission at $\lambda=3.75$ cm around the galactic center, obtained by Downes and Maxwell[11] with the Haystack 36-m telescope at a beam-width of 4'.2. Sgr A is the radio source at $(l, b) = (0, 0)$.

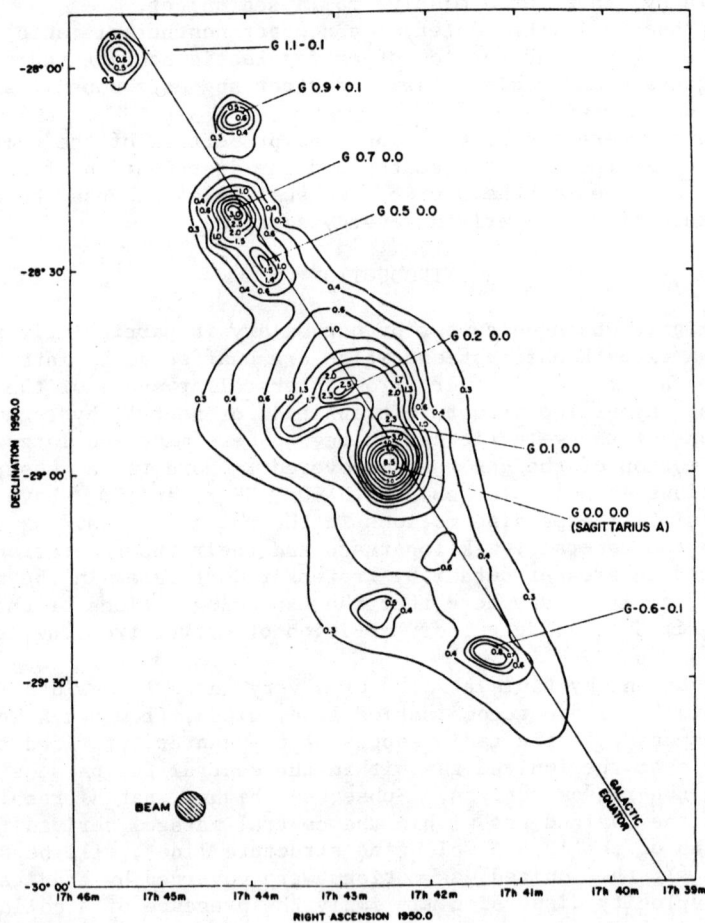

Fig. 1

Until observations able to resolve the structure of Sgr A were available, the nature of the radio source (e.g. thermal vs. nonthermal) had been ambiguous[12]. Suggestions by Lynden-Bell and Rees[13] that there might be a black hole in the galactic center as an energy source and that there might be a small diameter radio source associated with the surrounding active region motivated interferometric observations by Ekers and Lynden-Bell[14] and by Downes and Martin[8]. They identified several components; in particular, Downes and Martin identified 3 components: Sgr A West, Sgr A East and compact structure. Sgr A East was interpreted as a nonthermal source, mainly because of its spectrum, while Sgr A West and compact structure remained ambiguous in nature.

The nonthermal nature of the compact structure was definitively established when Balick and Brown[15] obtained an upper limit of 0".1 for the source size. This implied a source brightness temperature of $>10^7$ K.

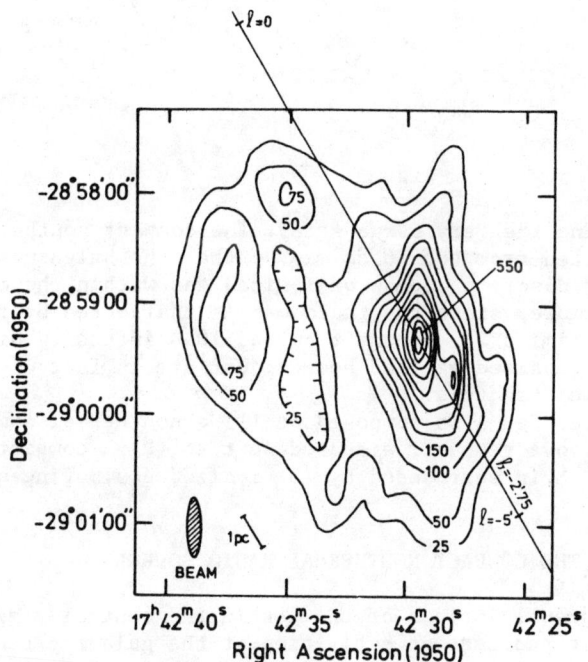

Fig. 2

An aperture synthesis map of Sgr A with angular resolution of 6".3 x 34" was first obtained by Ekers et al[16] (Figure 2) at $\lambda$ = 6 cm. It clearly shows Sgr A West and Sgr A East as spatially distinct features, while the compact nonthermal source is embedded in the middle of Sgr A West.

The thermal nature of Sgr A West is quite clear when one compares the radio brightness distribution at $\lambda$ = 6 cm with the intensity map of the 10 μ continuum[17], as shown in figure 3. The radio map of Sgr A West at a resolution of 2" x 8" was recently obtained by Brown, Johns-

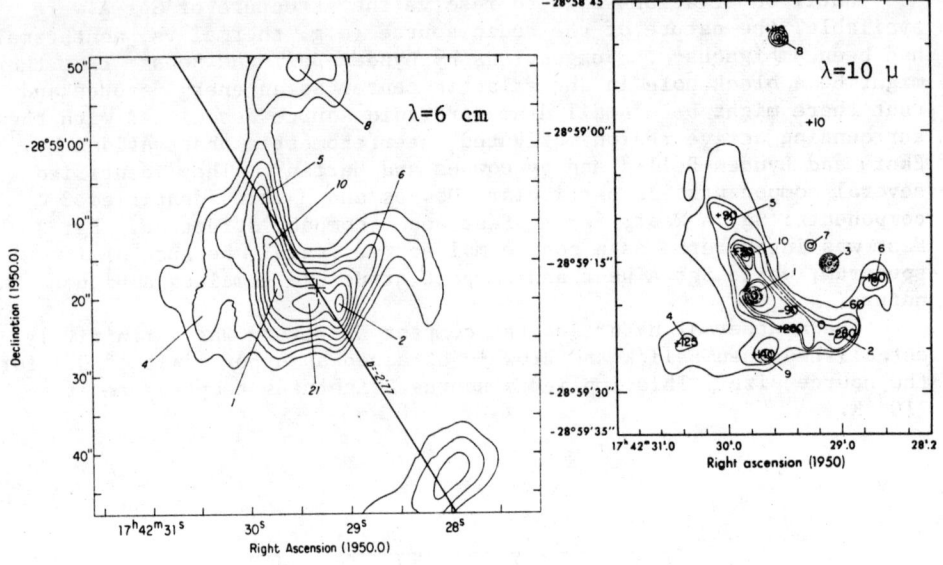

Fig. 3

ton and Lo[18] using the Very Large Array. The compact nonthermal source has been subtracted from the 6 cm map so that the brightness distribution is that of discrete clumps of ionized gas within the central parsec. Furthermore, since the radio map is unaffected by foreground extinction, the similarity of the 6 cm and 10 µ intensity distributions implies that the ionized gas and heated dust are in fact distributed along an arc-like structure.

To summarize, Sgr A is composed of (1) a nonthermal component, possibly a supernova remnant, situated next to (2) a compact nonthermal radio source which is surrounded by (3) ionized gas moving at high velocities.

## THE COMPACT NONTHERMAL RADIO SOURCE

With the unusual motions of the gas in the central region suggesting possible past and current activities at the galactic center, it is particularly intriguing to find there a compact nonthermal radio source with properties similar to those found in quasars and radio galaxies.

The properties of the radio source are summarized in Table I [19].

There is only an upper limit to the radio source size because the measured size is observed to vary as $\lambda^2$, where $\lambda$ is the observing wavelength[19], as shown in figure 4. This could be due to either interstellar electron scattering[20] or inhomogeneous source structure[21]. Kellermann et al[22] reported the detection of a 0".001 core, but other observations at different times using the same telescopes at the same wavelength gave conflicting results [23,24,19]. Recent $\lambda$=2.8 cm very long baseline interferometric (VLBI) observations, using the 5 times more sensitive Mk III recording system, on two separate occasions failed

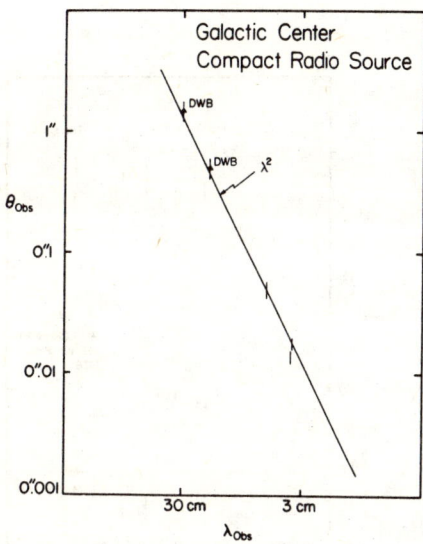

Fig. 4
DWB refers to measurements by Davies et al.[20]

to detect any core component stronger than 0.06 Jy (5σ with 90-second integration time)[25]. The results are, however, consistent with the $\lambda^2$-dependence of the observed size down to $\lambda=2.8$ cm.

To establish the intrinsic source size is urgent since it provides a more definitive constraint on source models[26]. This is best done at high frequencies ($\lambda \lesssim 1$ cm) to minimize both the interstellar scattering and optical depth effects. While mapping the brightness distribution of the source is obviously important, attempts to do so are complicated by (1) the faintness of the source, (2) the limited visibility of the source from the northern hemisphere, and (3) possible interstellar scattering.

Table I  Observed parameters of the compact source

| | |
|---|---|
| Scale Size | $<10^{15}$ cm |
| Radio luminosity | $3 \times 10^{33}$ erg/s |
| Brightness temperature | $>4 \times 10^9$ K |
| Spectral index | $\sim 0.2$ |
| Turnover frequency | $\gtrsim 22$ GHz |
| Flux variability $\Delta S/S_{min}$ | $<1$ over 7 years |
| Soft X-ray luminosity | $1.5 \times 10^{35}$ erg/s [27] |
| Hard X-ray luminosity | $<1.5 \times 10^{36}$ erg/s [28] |
| Upper limit to mass | $\lesssim 5 \times 10^6$ M$_\odot$ |

Since its first identification in 1974, the compact source at the galactic center has displayed no significant ($\Delta S/S_{min} > 1$) outbursts[29]. However, systematic flux monitoring has shown that variations in the flux density are present at various time scales as short as an hour[30]. Figure 5 shows the flux density measured at 2 frequencies simultaneously over a 2.5-year interval. At 2695 MHz ($\lambda=11$ cm), there was clearly a $\sim 10\%$ a year secular increase in flux density. Although a corresponding secular increase is not obvious at 8085 MHz ($\lambda=3.7$ cm), it could be masked by the larger fluctuation in the flux density. The secular increase in flux density can be explained by an expanding, optically thick, inhomogeneous source in which the optical depth is larger at longer wavelength. While the variation time scales may be used to set

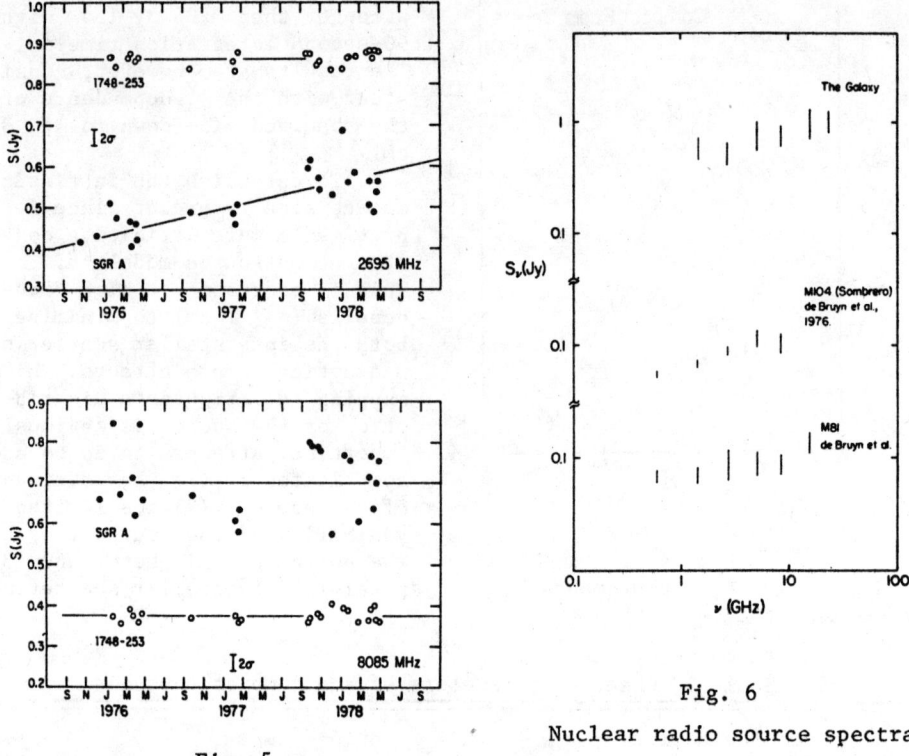

Fig. 5

Fig. 6
Nuclear radio source spectra

1748-253 is a reference source near the galactic center.

limits on the intrinsic size, the uncertainty of whether only a part of or the whole source is varying makes the result ambiguous.

The spectrum of the radio source may also provide information on the source structure[26]. Figure 6 shows the spectrum of the galactic center source along with the spectra of the compact radio sources in the nuclei of M81 and M104[31]. The slowly rising spectra are most simply interpreted as that of a self-absorbed incoherent synchrotron source. However, more definitive inferences of the source properties depend on two as yet unknown observables: the intrinsic source size and the turn-over frequency, the frequency above which the source becomes optically thin.

Polarization measurements would also be useful for defining further the properties of the radio source: If the apparent brightness temperature of $\sim 10^9$ K were intrinsic to the source, the radio emission might be due to gyrosynchrotron radiation by mildly relativistic electrons and significant circular polarization could be present[32]. However, inspection of existing measurements shows that circular polarization, if present, is less than a few percent[19].

To facilitate identification of the optical or infrared counterparts to the compact radio source at the galactic center, the absolute position is important and the latest measurement is at $\lambda=6$ cm[18]:

$$\alpha(1950) = 17^h42^m29\overset{s}{.}335 \pm 0.008, \quad \delta(1950) = -28°59'18\overset{"}{.}6 \pm 0\overset{"}{.}24 .$$

A detailed discussion of the nature of the compact source will be given by Rees[33]. However, empirical arguments may already exclude some models as possible energy source. A strong observational constraint is the small scale size $<10^{15}$ cm (100 a.u.). Pulsars and binary stellar radio sources[34] have very small scale sizes. Ordinary pulsars are ruled out because (1) the spectral index of pulsar radiation, $\alpha$, generally lies in the range $-3<\alpha<-1$ [35], whereas that of the compact radio source is 0.2, and (2) the most luminous pulsar known - the Crab pulsar - has an average radio luminosity of about 1000 times less. Binary stellar radio sources are unlikely to be the energy source for the galactic center source because (1) they are characterized by frequent, large outbursts[34]: $\Delta S/S_{min}$ is often 10 or more, and (2) the ratio of their X-ray to radio luminosities is at least 30 times larger than that of the galactic center source.

## FURTHER OBSERVATIONS

The unusual activities in the galactic center as suggested by anomalous motions of the gas and the recent discovery of the variable source of the $\gamma$-ray line emission make the deciphering of the nature of the compact radio source compelling.

To do so, more observations are needed: (1) determining the intrinsic size, the brightness distribution - to search for, for example, jet structure, (2) extending the source spectrum to millimeter wavelength, to define the turn-over frequency, (3) obtaining better limits on polarization, (3) monitoring variations in the source spectrum and studying very short time scale variation and scintillation effects. To decide whether the energy source of the compact radio source is massive or stellar, one would have to determine the mass of the object. One method is to determine the velocity dispersion of the central star cluster that the radio source may be associated with. Because of the very high extinction along the line of sight to the galactic center, this would have to be done at the near-infrared wavelength[36].

## REFERENCES

1. G. W. Rougoor, J. H. Oort, Proc. Nat. Acad. Sci. USA, 46, 91 (1960).
2. N. Z. Scoville, P. M. Solomon, P. Thaddeus, Ap. J., 172, 335 (1972).
3. J. Kaifu, T. Kato, T. Iguchi, Nature, Phys. Sci., 238, 105 (1972).
4. T. Bania, Ap. J., 216, 381 (1977).
5. J. H. Oort, this Proceeding (1982).
6. J. H. Oort, Ann. Rev. Astr. Ap., 15, 295 (1977).
7. T. Pauls, P. G. Mezger, E. Churchwell, Astr. Ap., 34, 327 (1974).

8.  D. Downes, A. H. M. Martin, Nature, 233, 112 (1971).
9.  J. H. Lacy, this Proceeding (1982).
10. J. H. Lacy, C. H. Townes, T. R. Geballe, D. J. Hollenbach, Ap. J., 241, 132 (1980).
11. D. Downes, A. Maxwell, Ap. J., 146, 653 (1966).
12. T. Jones, Astr. Ap., 30, 37 (1974).
13. D. Lynden-Bell, M. J. Rees, M.N.R.A.S., 152, 461 (1971).
14. R. D. Ekers, D. Lynden-Bell, Ap. Lett., 9, 189 (1971).
15. B. Balick, R. L. Brown, Ap. J., 194, 265, (1974).
16. R. D. Ekers, W. M. Goss, U. J. Schwarz, D. Downes, D. H. Rogstad, Astr. Ap., 43, 159 (1975).
17. E. Becklin, K. Matthews, G. Neugebauer, S. P. Willner, Ap. J., 219, 121 (1978).
18. R. L. Brown, K. J. Johnston, K. Y. Lo, Ap. J., 250, 155 (1981).
19. K. Y. Lo, M. H. Cohen, A. C. S. Readhead, D. C. Backer, Ap. J., 249, 504 (1981).
20. R. D. Davies, D. Walsh, R. Booth, M.N.R.A.S., 177, 319 (1976).
21. A. G. de Bruyn, Astr. Ap., 52, 439 (1976).
22. K. I. Kellermann, D. B. Shaffer, B. G. Clark, B. J. Geldzahler, Ap. J. (Letters), 214, L61 (1977).
23. B. J. Geldzahler, K. I. Kellermann, D. B. Shaffer, A. J., 84, 186 (1979).
24. K. Y. Lo, M. H. Cohen, R. T. Schilizzi, H. N. Ross, Ap. J., 218, 668 (1977).
25. K. Y. Lo, D. C. Backer, M. H. Cohen, in preparation (1982).
26. S. P. Reynolds, C. F. McKee, Ap. J., 239, 893 (1980).
27. M. G. Watson, R. Willingale, J. E. Grindlay, P. Hertz, Ap. J., 250, 142 (1981).
28. R. G. Cruddace, G. Fritz, S. Shulman, H. Friedman, J. McKee, M. Johnson, Ap. J. (Letters), 222, L95 (1978).
29. R. D. Ekers, Highlights Astr., 5, 143 (1980).
30. R. L. Brown, K. Y. Lo, Ap. J., in press (1982).
31. A. G. de Bruyn, P. C. Crane, R. M. Price, J. B. Carlson, Astr. Ap., 46, 243 (1976).
32. R. M. Hjellming, D. M. Gibson, IAU Symposium 86, Radio Physics of the Sun, ed. M. R. Kundu, T. E. Gergely (Reidel: Dordrecht, 1980), p. 209.
33. M. J. Rees, this Proceeding (1982).
34. D. Gibson, Ph.D. thesis, University of Virginia (1976).
35. W. Sieber, Astr. Ap., 28, 237 (1973).
36. K. Y. Lo, J. H. Lacy, T. R. Geballe, S. E. Persson, this Proceeding (1982).

# MOLECULAR GAS NEAR THE GALACTIC CENTER
## - RECENT CONTRIBUTIONS FROM RADIOASTRONOMY

Rolf Güsten
Max-Planck-Institut für Radioastronomie, 5300 Bonn, F.R.G.

### INTRODUCTION

To improve our knowledge about the physical nature of the molecular clouds close to the galactic center, several high resolution studies with the Effelsberg radio telescope have been performed during the last year. Only a preliminary summary of the results will be given here.

Based on the 6 cm $H_2CO$ distribution[1], parts of the inner 15 arc min of the nucleus have been mapped in the two lowest inversion lines of ammonia ($NH_3(J,K)$, $J = K = 1,2$)[2], and in the $2_{11}-2_{12}$ transition of formaldehyde[3]. Toward selected positions, the higher transitions $NH_3(4,4)$ and $(5,5)$, with excitations of 200 and 300 K above the ground level, have also been measured. Results of a sensitive search for water masers ($H_2O$) in the center region are reported. Finally, recent detections of "high-velocity" gas are summarized.

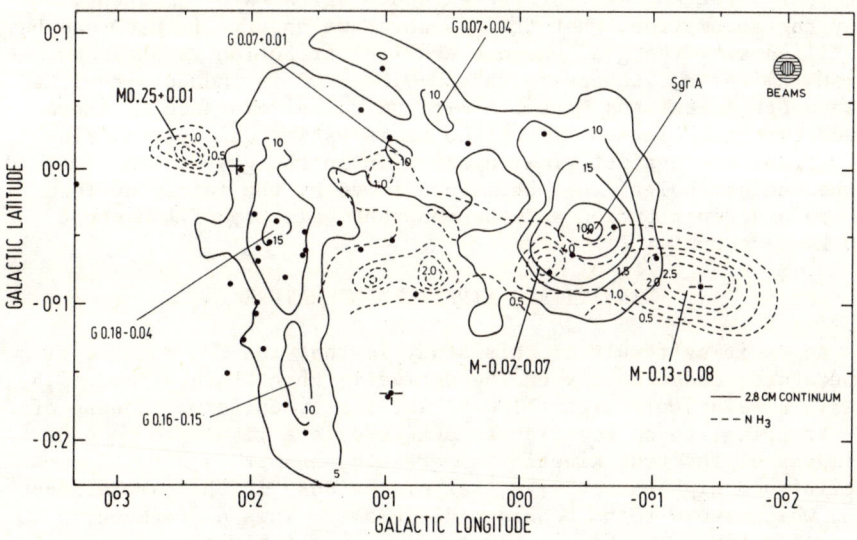

Fig. 1. Overlay of the ammonia distribution $T_A(NH_3(1,1))$[2] (dashed contours) on the 2.8 cm radio continuum emission[8]. The ammonia peaks are labelled with M (for molecular cloud) followed by their $(l,b)$-coordinates. Dots represent compact radio continuum sources[5], crosses mark the positions of the new $H_2O$ masers[4].

## GLOBAL PROPERTIES OF THE CLOUDS

The high spatial resolution of $\sim$2 pc (40") at the frequencies of the ammonia lines ($\nu$ $\sim$24 GHz) shows a considerable amount of structure not apparent in previous maps made with typically 2-3' beams. Several ammonia peaks have been identified, of which some are still unresolved. These peaks are probably mass condensations rather than hot spots as they also show up in the 2 cm formaldehyde line, which mainly traces high-density regions.

The most prominent peak is M-0.13-0.08, also known as the "+20 km s$^{-1}$ cloud", for which the regular rotation pattern ($\sim$5 km s$^{-1}$ arc min$^{-1}$) as well as the derived H$_2$-density both indicate a total mass of $\sim 10^6$ M$_\odot$. The linewidth as seen with our 40" beam is typically $\sim$15 km s$^{-1}$, except for M-0.02-0.07, often called the "+40 km s$^{-1}$ cloud". This is the condensation closest (in projection) to the galactic center, and is unusual in showing a linewidth $\Delta v$ $\sim$25 km s$^{-1}$. The central velocity, however, is roughly constant (50±5 km s$^{-1}$) over the whole complex of clouds between $-2' \leq l \leq 15'$, perhaps indicating some physical relation.

While M-0.13-0.08 is likely to be in a foreground position, because it seems to account for a deep depression of the IR-radiation of the central stellar bulge, the location of the "+40 km s$^{-1}$ cloud" relative to the galactic center has been subject of much discussion. In 2 cm H$_2$CO we observe absorption against the Sgr A East shell, but not against the extended Sgr A halo component. Under the assumption, that the gas which we observe in H$_2$CO can be identified with the gas, whose overall distribution is shown by the NH$_3$ emission, it appears that M-0.02-0.07 is indeed sandwiched between Sgr A West and East[1]. A minimum radial distance of these clouds to the nucleus of order 100 pc is estimated[1] by requiring their stability against tidal disruption in the gravitational field of the nuclear bulge. This seems confirmed by the fairly uniform temperature distribution and the constant velocity of the whole "+50 km s$^{-1}$ complex".

## DENSITY AND KINETIC TEMPERATURE

An exciting result of this study is that the NH$_3$ rotational temperature, based mainly on the optically thin NH$_3$(4,4) and (5,5) lines, is relatively high ($\sim$100 K) and fairly uniform. Because of line-trapping, these rotation temperatures are likely to be underestimates of the true kinetic temperature, and it is hard to reconcile such high values with that of the dust in the same direction, which seems to be considerably cooler. This difference in the temperatures of the gas and the dust may imply some direct heating source of the gas, such as cosmic rays.

H$_2$ densities of $\sim 10^5$ cm$^{-3}$ for the main condensations have been derived by comparing the H$_2$CO 6 cm and 2 cm transitions in a LVG radiative transfer program[3]. Such high densities make thermalization of the ammonia doublets likely. For $T_{ex} = T_{kin}$, the observed line temperatures correspond to a beam filling factor of $\sim$0.2, indicating clumping on even smaller scales.

## STAR FORMATION

To verify if star formation is going on in the galactic center clouds, a sensitive search (5 Jy-limit) for $H_2O$ masers has been made[4] toward all of the compact radio continuum sources[5] and the region of the molecular clouds mapped in $NH_3$. Only three weak sources were detected (see Fig. 1), two of which appear to be associated with the center clouds in both position and velocity. Normally, the masers are signposts of stars more massive than $\sim 10\ M_\odot$ during their earliest ($\lesssim 5\ 10^5$ yrs) phase of evolution. If we attribute the high thermal flux of the galactic center region to the ionizing radiation of massive stars, a ten times higher number of masers is expected than actually observed. Thus either the stars are quite evolved and star formation is not a continuous process, or the Initial Mass Function is deficient in stars more massive than $\sim 20\ M_\odot$.

## HIGH-VELOCITY GAS

Rapid motion in the inner nucleus has been noted since 1974, when a broad hydrogen radio recombination line was detected toward Sgr A. Recent measurements of this broad recombination line at 14.7 GHz[6] with higher angular resolution (55") confirm linewidths of the order 200 km s$^{-1}$, and a definite velocity gradient in the sense of galactic rotation is seen. This is different from the high-velocity motion of the neutral gas as detected toward Sgr A[7] in $H_2CO$ and HI absorption extending to -210 km s$^{-1}$. In these clouds equilibrium rotation is ruled out, and expansion dominates the profiles, perhaps indicating recent ($\lesssim 10^6$ yrs) gas ejection out of the nucleus.

## REFERENCES

[1] Güsten, R., Downes, D.: 1980, Astron. Astrophys. **87**, 6
[2] Güsten, R., Walmsley, C. M., Pauls, T.: 1981, Astron. Astrophys. **103**, 197
[3] Güsten, R., Henkel, C.: 1982, in preparation
[4] Güsten, R., Downes, D.: 1982, in preparation
[5] Downes, D., Goss, W. A., Schwarz, U. J., Wouterloot, J. G. A.: 1978, Astron. Astrophys. Suppl. **35**, 1
[6] Pauls, T., Mezger, P. G.: 1982, in preparation
[7] Güsten, R., Downes, D.: 1981, Astron. Astrophys. **99**, 27
[8] Pauls, T., Downes, D., Mezger, P. G.: 1976, Astron. Astrophys. **46**, 407

## THE RELATIVE LOCATIONS OF THE SGR A
## MOLECULAR CLOUDS AND CONTINUUM SOURCES

Aa. Sandqvist
Stockholm Observatory, S-133 00 Saltsjöbaden, Sweden

### ABSTRACT

Formaldehyde has been detected directly towards Sgr A West in the 2 mm, $J = 2_{12} - 1_{11}$ emission line with a 1' resolution. The formaldehyde is located near the galactic center in a molecular belt which contains two major concentrations with radial velocities of +50 and +15 km s$^{-1}$, respectively, and which displays a mean velocity gradient of 6.7 km s$^{-1}$ arcmin$^{-1}$ in the same sense as that of galactic rotation. The continuum radio source Sgr A East is probably on the near side of the molecular region, while Sgr A West appears to be on the far side implying that the molecular region has a velocity component directed in towards the galactic center.

### INTRODUCTION

The molecular complex associated with Sgr A, the strong radio source near the galactic center, has been studied in detail in the absorption lines of H I, OH and $H_2CO$ (6 cm)$_*$. Through a series of lunar occultations[1], observed with the NRAO 43 m radio telescope in 1968-1970, the continuum source Sgr A was resolved into several components, now generally known as Sgr A West and Sgr A East. Furthermore, the molecular cloud complex was also resolved into several components, with major features having radial velocities of +50 and +25 km s$^{-1}$. The lunar occultations showed evidence of a general velocity gradient across the complex. Basing my conclusions on the analysis and results of the occultations, I suggested that the molecular complex lies in front of Sgr A West and therefore has a velocity component directed in towards the galactic center at the same time that it partakes in a rotation about the center of the Galaxy. Furthermore, I showed that at least part (and possibly all) of Sgr A East lies in front of the molecular complex and is therefore not absorbed by this region.

It is noteworthy that the OH observations discussed above show a concentration of molecules directly towards Sgr A West that is smaller than in the surrounding regions. This effect was also noted in aperture synthesis observations[2] of 6 cm $H_2CO$ absorption in the Sgr A complex and was interpreted as being due to the molecular region being behind the continuum sources and therefore moving out from the galactic center. An active debate, as to where this gas really is, has been taking place in the literature since 1974.

In the analysis of the occultation and aperture synthesis absorption data, it was assumed that all the continuum sources lay be-

---

*The National Radio Astronomy Observatory is operated by Associated Universities Inc., under contract with the National Science Foundation.

0094-243X/82/830012-06$3.00 Copyright 1982 American Institute of Physics

hind the molecular region. Absorption line data are affected by the relative locations of the continuum and the molecular regions, emission line data on the other hand are not. If we therefore abandon the above assumption and reanalyze the absorption data, making a comparison with an emission map of the molecular region, using e.g. the 2 mm $H_2CO$ emission line, we may be able to get some indication of the relative location of the molecular region and the continuum sources.

## OBSERVATIONS AND RESULTS

The 2 mm $H_2CO$ emission in the Sgr A molecular cloud complex was mapped in the 140.8 GHz, $J = 2_{12} - 1_{11}$ transition using the 11 m telescope at Kitt Peak during two observing sessions in 1979 and 1980. The NRAO 2 mm uncooled receiver ($\sim$ 4700 K) was used during the first session, but most of the observations reported herein were made a year later with a cooled ($\sim$ 2000 K) version of the above receiver. The 0.5 and 1.0 MHz, 256 channel filter bank receivers yielded velocity resolutions of 1.1 and 2.1 km s$^{-1}$, respectively. The beamwidth was 1' and a map was produced consisting of grid points with 1' spacing. The map is centered on the position of Sgr A West at $\alpha(1950.0) = 17^h42^m29^s.3$, $\delta(1950.0) = -28°59'18"$ (the reference "off" point in the position switching observing mode was $\alpha(1950.0) = 17^h41^m29^s.3$, $\delta(1950.0) = -28°59'18"$). The temperature calibration was made using the standard chopper wheel method[3].

A region of about 6' x 13' was mapped yielding profiles with a maximum value of $T_A^*$ reaching up to 1.4 K and half power velocity widths ($\Delta V$) varying from about 30 km s$^{-1}$ in the southwest region to about 50 km s$^{-1}$ in the northeast part of the complex. One of the most important of the profiles is of course the one observed directly towards Sgr A West and this is presented in Figure 1. The integrated profile ($\int_{line} T_A^* dV$) map is shown in Figure 2 which also contains isovelocity contours of profile velocity at maximum $T_A^*$.

## DISCUSSION

The 2 mm $H_2CO$ profile in Figure 1 shows that there is definitely formaldehyde seen towards Sgr A West in the molecular complex near the galactic center. The profile, although relatively weak ($T_A^* \sim 0.4$ K), is broad ($\Delta V \sim 50$ km s$^{-1}$) and centered on a radial velocity with respect to the local standard of rest of $\sim$ 40 km s$^{-1}$. A similar profile has been observed[4] directly towards Sgr A West with approximately the same beamwidth in the 267.6 GHz, $J = 3-2$ transition of $HCO^+$ using the University of Texas MWO mm-wave telescope. The question remains, however, whether the molecular gas is on the near or far side of the galactic center.

The formaldehyde, as seen in the 2 mm map in Figure 2, appears to be concentrated in two major components, one in the northeastern part of the complex, the other in the southwestern part. The northeast component has a radial velocity of +50 km s$^{-1}$ while the south-

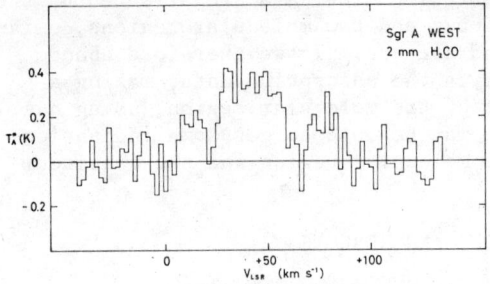

Fig. 1. The 2 mm $H_2CO$ emission profile observed directly towards Sgr A West with a 1' beam. The velocity resolution is 2.1 km s$^{-1}$.

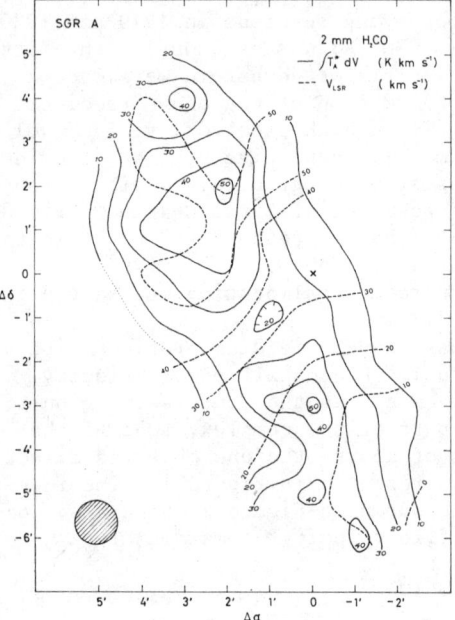

Fig. 2. Contour map of the formaldehyde distribution near the galactic center. The solid lines represent values of $\int_{line} T_A^* dV$ in K km s$^{-1}$ for the 2 mm $H_2CO$ emission line, while the broken lines represent the radial velocity with respect to the local standard of rest of the maximum $T_A^*$. The cross is the origin of the coordinate system and is the position of Sgr A West $\alpha(1950.0) = 17^h42^m29\overset{s}{.}3, \delta(1950.0) = -28°59'18"$. The beam is shown in the lower left corner.

west component peaks up at +15 km s$^{-1}$. There is a significant velocity gradient across the complex, in confirmation of the occultation results. The large velocity gradient, which is not linear but has a mean dimension of $\sim$ 6.7 km s$^{-1}$ arcmin$^{-1}$, ends near the maximum of the +50 km s$^{-1}$ component. The components appear to be immersed in a general molecular belt (defined by the two outermost contours), reaching from the southwest to the northeast of the map and probably continuing outside the range of the map. The two components could of course be far removed from one another, on opposite sides of the galactic center, for example. But the relatively constant width of the belt from southwest to northeast and the smooth velocity gradient seem to imply some physical connection between the two components. For the purpose of the remainder of this paper, I shall therefore assume that the +50 and +15 km s$^{-1}$ clouds are components of a continuous belt of molecular gas. I shall furthermore restrict my reanalysis of the absorption data to moving the major continuum sources relative to this belt.

Four different maps of the total optical depth ($\tau_6$), calculated from 6 cm $H_2CO$ aperture synthesis absorption data[2], have been produced and are presented in Figures 3(a) - (d). It is assumed that the continuum source, Sgr A, is made up of two sources, Sgr A West (W) and Sgr A East (E), superimposed on an extended non-varying background (B), which is defined by the outermost contour in Figure 1 of reference 2. B is assumed to be behind the molecular complex in all four cases. The assumptions involving B are rather uncertain, but necessary since some continuum radiation must exist on the far side of the molecular complex to produce absorption. To obtain the total optical depth, integrated across the velocity range covered in Figures 2(a) and (b) in reference 2, the 6 cm $H_2CO$ brightness temperature maps were added together in the velocity ranges -25.2 to -12.8 km s$^{-1}$ and +9.0 to +80.4 km s$^{-1}$. Linear interpolation was performed over the 0 km s$^{-1}$ feature, stretching over the maps in the velocity range of -9.7 km s$^{-1}$ to +5.9 km s$^{-1}$, since this feature is believed to be a foreground object with respect to the Sgr A complex[5]. Maps of $\tau_6$ were then produced with the four possible different locations of E and W with respect to the molecular region: Figure 3(a) - Sgr A East and Sgr A West both behind the molecular region; Figure 3(b) - Sgr A East and Sgr A West both in front of the molecular region; Figure 3(c) - Sgr A East behind and Sgr A West in front of the molecular region; Figure 3(d) - Sgr A East in front of and Sgr A West behind the molecular region. For ease in comparisons, the 2 mm $H_2CO$ map of the relevant region, taken from Figure 2, is superimposed as a broken line contour map on the $\tau_6$ maps in Figure 3.

It is apparently in Figure 3(d) that the 6 cm absorption results best resemble the 2 mm emission results. One should not attempt to make a detailed comparison since the 6 cm data was taken with a resolution of 20" x 40" whereas the resolution of the 2 mm data is 1'. However, certain "problems" are rather obvious in Figures 3(a) - 3(c). In Figure 3(a) there is a depression in the 6 cm distribution just where the 2 mm distribution rises towards a maximum. This discrepancy occurs at the position of Sgr A East, implying that the assumption, that Sgr A East is on the far side of the molecular region, is erroneous. In Figure 3(b) it is assumed that Sgr A East is on the near side of the molecular region, and indeed the 6 cm map bears a closer resemblance to the 2 mm map in the region of Sgr A East. However, the assumption in this figure, that Sgr A West is also on the near side of the molecular region, has produced two strong 6 cm concentrations near the position of Sgr A West that have no counterpart in the 2 mm map. This implies that the assumption that Sgr A West is on the near side of the molecular region is erroneous. The worst possible situation is shown in Figure 3(c) where both the erroneous assumptions have been combined. Figure 3(d) is thus considered to contain the best agreement between the 6 cm absorption and the 2 mm emission $H_2CO$ data. That the 6 cm map appears to show

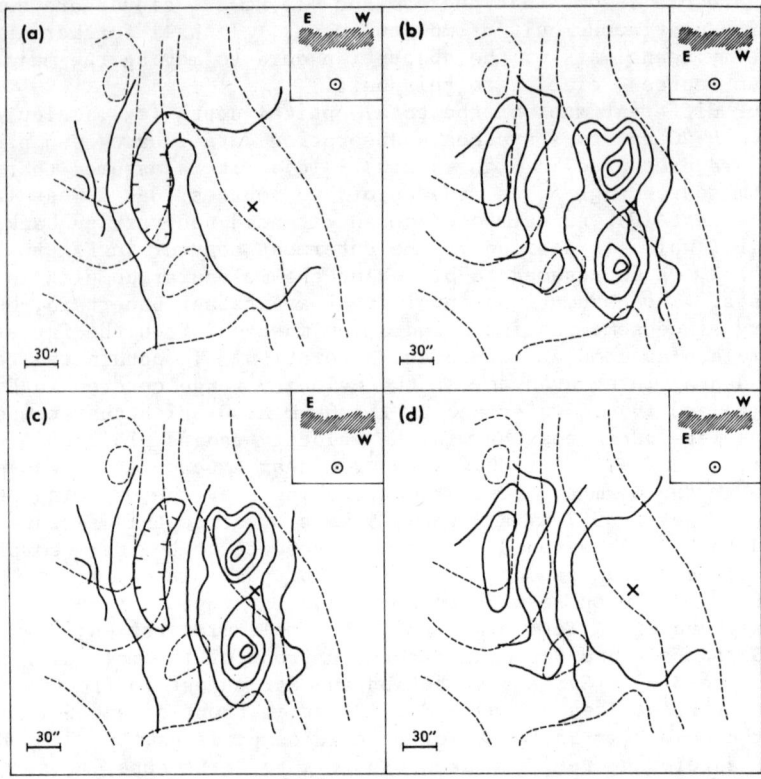

Fig. 3. Total optical depth (solid contours) of the 6 cm $H_2CO$ absorption line integrated over the velocity range discussed in the text. The optical depth has been calculated for four different assumptions (a-d) concerning the relative locations (shown in the upper right inserts) of the continuum sources (E - Sgr A East, W - Sgr A West) and the molecular region (shaded region). The sun is represented by the symbol "$\odot$". The dashed contours are the relevant parts of the 2 mm $H_2CO$ $\int T_A dV$ map from Figure 2. The cross marks the position of Sgr A West.

a slight underabundance of molecules at the position of Sgr A West may be due to the assumption involving an incorrect level of the background radiation (B). It therefore appears that Sgr A East is on the near side of the molecular region and Sgr A West on the far side. This would confirm the conclusions drawn from the occultation results and would imply that the molecular region is moving in towards the galactic center while partaking in a rotation about the center of the Galaxy.

## CONCLUSION

Emission in the 2 mm $H_2CO$ line has been observed towards Sgr A West. The emission comes from a molecular belt near the galactic center which has concentrations at velocities of +50 and +15 km s$^{-1}$ and which displays a large velocity gradient in the same sense as that of galactic rotation. Evidence is presented that the continuum radio source Sgr A East lies on the near side of the molecular region, while Sgr A West lies on the far side. This would imply that the molecular region has a velocity component directed in towards the galactic center.

This research has been supported by the Swedish Natural Science Research Council.

## REFERENCES

1. Aa. Sandqvist, Astron. Astrophys. 33, 413 (1974).
2. J.B. Whiteoak, D.H. Rogstad, I.A. Lockhart, Astron. Astrophys. 36, 245 (1974).
3. B.L. Ulich, R.W. Haas, Astrophys. J. Suppl. 30, 247 (1976).
4. A. Wootten, Aa. Sandqvist, R.B. Loren, (in preparation).
5. Aa. Sandqvist, Astron. J. 75, 135 (1970).

# DEPRESSION OF MOLECULAR EMISSION IN THE LINE OF SIGHT OF SGR A WEST

Y. Fukui, H. Ogawa
Department of Astrophysics, Nagoya University
Nagoya, Japan

S. Deguchi
Five College Radio Astronomy Observatory
University of Massachusetts
MA 01003, U.S.A.

H. Suzuki
Department of Physics, Kyoto University
Kyoto, Japan

ABSTRACT

The galactic center region has been mapped in the 4-mm emission lines of HCCCN (J=8-7) and $H_2CO$ (JK-K+=$1_{01}$-$0_{00}$) with a 1.5 arc min beam. The molecular lines are found to show depression in intensity in the line of sight of Sgr A West. Comparison with other molecular data indicates that $NH_3$ also shows a significant depression while HCN and $HCO^+$ show little sign of similar depression. Based on some density estimates we suggest that the depression means abnormally reduced abundance in HCCCN, $NH_3$, and $H_2CO$ in the line of sight of Sgr A West. The difference in the degree of depression could be interpreted in terms of a time-dependent ion-molecule reaction scheme because $HCO^+$ and HCN are formed much more rapidly than the other molecules in the scheme.

INTRODUCTION

Toward the galactic nucleus, Sgr A West, various molecular emission lines are observed at 30-50 km/s $v_{LSR}$. In the 3.4-mm emission lines of HCN and $HCO^+$ we mapped the region with 2 arc min resolution by using the 6 m telescope at Tokyo and found that these molecules emit rather strong radiation in the line of sight of Sgr A West (Fukui et al. 1977; Paper I, Fukui et al. 1980; Paper II). Recently at Bonn $NH_3$ emission was mapped in this region with 40 arc sec beam size and lack of $NH_3$ toward Sgr A West was revealed (Gusten et al. 1981). In this paper we describe the results of mapping observations in the 4-mm emission lines of HCCCN and $H_2CO$, and discuss the implications of molecular depression in the line of sight of Sgr A West.

OBSERVATIONS AND RESULTS

Observations were made in September 1981 by using the NRAO 36 ft telescope at Kitt Peak. We employed the dual channel cooled mixer,

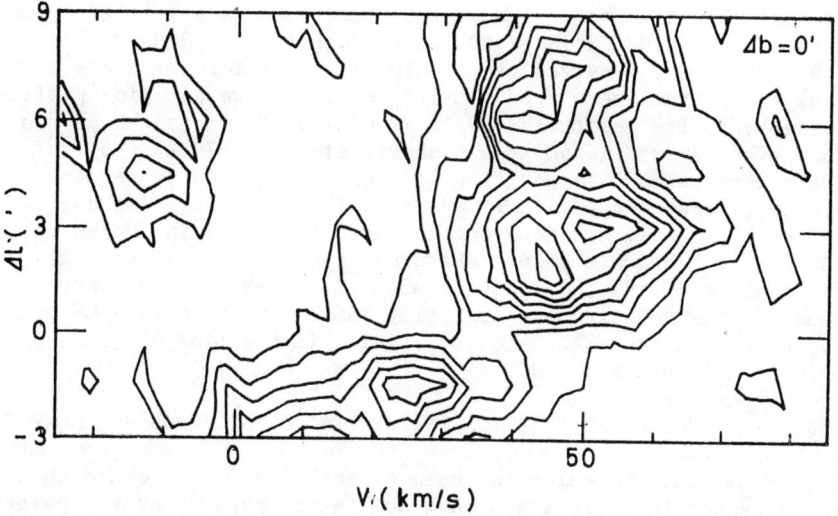

Figure 1.a Longitude-velocity diagram of 8-7 transition of HCCCN. Offsets are given by taking Sgr A West as the origin. Contour interval is 0.1 K $T_A^*$.

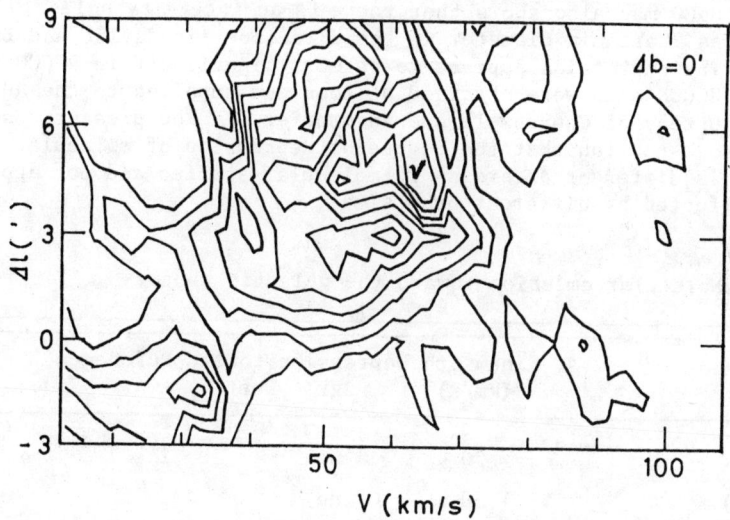

Figure 1.b Longitude-velocity diagram of $1_{01}$-$0_{00}$ transition of $H_2CO$. Contour interval is 0.1 K $T_A^*$.

the typical system temperature of which was about 500 K SSB for each orthogonal polarization.  1-MHz resolution 128-channel filterbanks were used and the HCCCN and $H_2CO$ transitions (72,783.8 MHz and 72,838.0 MHz) were observed simultaneously.  As for the HCCCN line 250-kHz filterbanks were used simultaneously and we got line profiles of better velocity resolution.  At 72 GHz the 36 ft telescope has a 1.5 arc sec beam size and the beam efficiency is 70 %.  All of the data were taken in a position switching mode with 30 sec intervals with reference position 1° apart in the same elevation angle.

Observations were done along the galactic plane including Sgr A West with 1.5 arc min grid.  Figure 1 shows the longitude-velocity diagrams of the two lines at $-3' \leq \Delta\ell \leq +9'$.  The distributions of peak velocity are basically similar to those of HCN and $HCO^+$, while HCCCN and $H_2CO$ show depression in the line of sight of Sgr A West unlike HCN and $HCO^+$.

DEPRESSION OF MOLECULAR EMISSION IN THE LINE OF SIGHT OF SGR A WEST

We summarize the molecular data of emission lines toward the galactic center in table 1 and show some strip maps along the galactic plane near Sgr A West in figure 2.  From these compilations we notice that HCN and $HCO^+$ show little intensity depression in the line of sight of Sgr A West and HCCCN, $NH_3$, and $H_2CO$ show significant depression.  It is also noticed that molecules depleted toward Sgr A West have small linewidth of about 20 km/s much smaller than HCN and $HCO^+$.  The data were taken with 1.5-2.0 arc min beam size except $NH_3$ and the difference cannot be due to different beam size. The 2-mm $H_2CO$ map also shows that there is an intensity hole as large as 2 arc min diameter in the direction (Sandqvist and Bernes 1980).  The depression appears to be most significant in HCCCN. (The 1-0 HCCCN data were obtained at Bonn and supplied to the authors by the courtesy of Churchwell and Angerhofer.)  The present results appear to establish that the degree of depression of molecular emission is different according to molecular species and not appreciably affected by different beam sizes.

Table I  Molecular emission toward the galactic center

| Molecule | Linewidth (km/s) | Depression toward Sgr A West | Reference |
| --- | --- | --- | --- |
| HCN(1-0) | ~ 50 | no | 1 |
| $HCO^+$(1-0) | ~ 50 | no | 2 |
| HCCCN(8-7,1-0) | ~ 20 | yes | present work, 5 |
| $H_2CO(1_{01}-0_{00}, 2_{12}-1_{11})$ | ~ 20 | yes | present work, 4 |
| $NH_3(1,1),(2,2)$ | ~ 15 | yes | 3 |

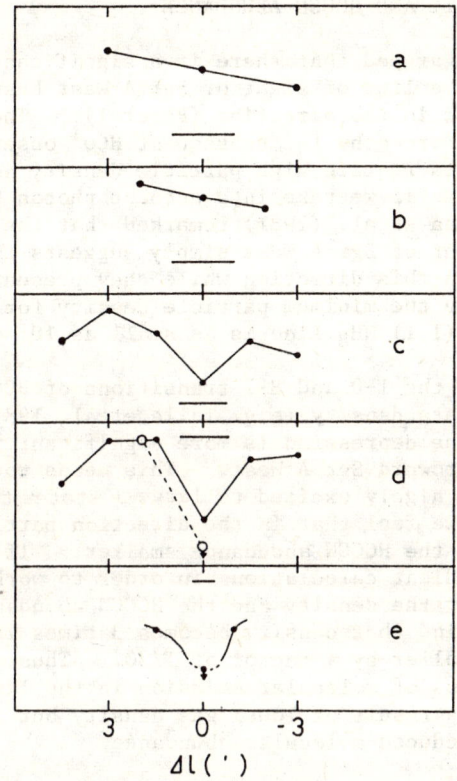

Figure 2. Distribution of molecular emission in arbitrary scale toward Sgr A West. Beam sizes are shown by horizontal bars.

a HCN(Paper I), b HCO$^+$(Paper II), c H$_2$CO(present work), d HCCCN(8-7) (present work) and HCCCN(1-0, F=2-1)(Ref. 5). The 1-0 data are shown by open circles, and the plot at Sgr A West is an upper limit read by the authors from the line profile, and e NH$_3$(Ref. 3). The plot at Sgr A West is an upper limit given in the reference.

As for the more abundant CO molecule a moderate intensity decrease of about 20 % toward Sgr A West was reported (Liszt et al. 1975). The CO emission has about 50-km/s linewidth and appears to be of a group with HCN and $HCO^+$.

## ESTIMATES OF DENSITY AND HCCCN ABUNDANCE

Previously we argued that there is a significant amount of molecular gas in the line of sight of Sgr A West based on rather strong HCN emission in the direction (Paper I). The suggestion appears to be reinforced by the subsequent $HCO^+$ observations (Paper II). The two lines require high particle density as large as at least $10^4$ $cm^{-3}$ even if we take into account photon trapping. On the contrary, Gusten et al. (1981) remarked that the lack of $NH_3$ in the line of sight of Sgr A West simply suggests that there is no high density gas in this direction while they presented no density estimate. We note the minimum particle density for collisional excitation of the (1,1) $NH_3$ line is as small as $10^3$ $cm^{-3}$.

Comparison of the 1-0 and 8-7 transitions of HCCCN may give some clue to estimate density (e.g. Tolle et al. 1981). We see in figure 2 that the depression is more significant in the 1-0 line than the 8-7 line toward Sgr A West. This means toward Sgr A West HCCCN is more highly excited to large J states than the surroundings. Therefore we feel that in the direction particle density becomes larger and the HCCCN abundance smaller. If we employ the large velocity gradient calculations in order to work out the line temperature, we get the density and the HCCCN abundance as compiled in table 2. We find that density becomes 3 times larger and the HCCCN abundance smaller by a factor of 1/30. Thus we think that the lack or weakness of molecular emission in the line of sight of Sgr A West is not a result of lower gas density but a manifestation of significantly reduced molecular abundance.

It is interesting to ask if the derived HCCCN abundance is in a range typical to galactic molecular clouds or not. Fractional abundance of various molecules decreases with density in dense interstellar clouds (e.g. Wootten et al. 1978), and we may need some reference source in order to make a comparison. Here we pick up TMC1 simply because we have the 8-7 data of HCCCN of this cloud and the 1-0 data taken at Bonn is available (Walmsley et al. 1980). Thus we get the same combination of telescopes and the same transitions as the galactic center. Results of a similar calculation are presented in table 3. The HCCCN abundance in TMC1 is larger than that toward Sgr A West and at the density around $10^4$ $cm^{-3}$ we find no evidence of sharp change in HCCCN abundance. Thus we suggest that the HCCCN abundance toward Sgr A West is significantly depleted as compared with TMC1.

Table II  Density and HCCCN abundance in the galactic center region

| Position | $n(H_2)$ | $n(HC_3N)/n(H_2)$ |
|---|---|---|
| (0',0') | $1.1 \times 10^4$ cm$^{-3}$ | $5 \times 10^{-11}$ |
| (2',0') | $3.5 \times 10^3$ | $1.5 \times 10^{-9}$ |

Notes; Position is given in the offsets from Sgr A West in (l,b). The velocity gradient, R/V, was assumed to be 2 pc/km/s, and $T_k$ 50K.


Table III  Density and HCCCN abundance in TMC1

| Position | $n(H_2)$ | $n(HC_3N)/n(H_2)$ |
|---|---|---|
| 2E, 0S  in arc min | $1.0 \times 10^4$ cm$^{-3}$ | $2 \times 10^{-9}$ |
| 0E, 2N | 1.3 | 3 |
| 2W, 4N | 1.8 | 1.5 |
| 4W, 6N | 1.8 | 1.5 |
| 6W, 8N | 3.5 | $3 \times 10^{-10}$ |

Notes; Position is given in the offsets from HCCCN (1-0) peak (Ref. 7) in R.A. and decl.  The velocity gradient was assumed to be 2 pc/km/s, and $T_k$ 10 K.

SPECULATIONS

In the above argument, abnormally reduced abundance in some molecules is suggested, and we may speculate about the origin of the anomaly. A staightforward way is an explanation by means of differential photo-dissociation due to strong radiation field of the galactic nucleus. But we find it unfeasible because there is no clear separation in photo-chemical properties among the observed molecules. Alternatively, the anomaly could be expalined if the chemical process in the cloud is highly time dependent, i.e. the cloud is very young with an age of about $10^4$ yr. A time-dependent ion-molecule reaction scheme indicates that $HCO^+$ and HCN are formed much more rapidly than HCCCN and $NH_3$ and to a lesser extent than $H_2CO$ (Suzuki 1979).

A fuller account of this work will be published elsewhere. We thank R. Kawabe for large velocity gradient calculations of HCCCN. Thanks are also due to Ed. Churchwell and P. Angerhofer who kindly made their 1-0 HCCCN data available to the authors prior to publication.

REFERENCES

1. Y. Fukui, T. Iguchi, N. Kaifu, Y. Chikada, M. Morimoto, K. Nagane, K. Miyazawa, and T. Miyaji, Publ. Astron. Soc. Japan 29, 643 (1977), Paper I
2. Y. Fukui, N. Kaifu, M. Morimoto, and T. Miyaji, Ap. J. 241, 147 (1980), Paper II
3. R. Gusten, C. M. Walmsley, and T. Pauls, A. Ap. (1981), in press
4. Aa. Sandqvist and C. Bernes, IAU Symp. No.87, p.103 (1980)
5. Ed. Churchwell and P. Angerhofer, private communication.
6. H. S. Liszt, R. H. Sanders, and W. B. Burton, Ap. J. 198, 537 (1975)
7. F. Tolle, H. Ungerechts, C. M. Walmsley, G. Winnewisser, and E. Churchwell, A. Ap. 95, 143 (1981)
8. A. Wootten, N. J. Evans II, R. Snell, and P. Vanden Bout, Ap. J. (Letters) 225, L143 (1978)
9. C. M. Walmsley, G. Winnewisser, and F. Toelle, A. Ap. 81, 245, (1980)
10. H. Suzuki, Prog. Theor. Phys. Kyoto 62, 937 (1979) and IAU Symp. No.87, p. 337 (1980)

Chapter II    Infrared Continuum Observations

## INFRARED OBSERVATIONS OF THE GALACTIC CENTER

Ian Gatley
United Kingdom Infrared Telescope

### ABSTRACT

Observations of the thermal emission from dust grains in the inner Galaxy are used to determine the total bolometric luminosity of the central parsec; a value in the range of $1 - 3 \times 10^7$ $L_\odot$ is obtained. Much of this luminosity originates in the source which ionises the gas in the HII region Sgr A. The dust density in the central parsec is very low, providing much less than one magnitude of visual extinction across this volume. Maps of infrared color temperature show that the grains are heated by a luminosity source which is centrally concentrated about the position of the galactic nucleus.

### INTRODUCTION

The advantages of infrared observations of the Galactic center follow simply from an important property of interstellar dust: the absorption efficiency decreases with increasing wavelength.[1] This has two immediate consequences:
1. At near infrared wavelengths the extinction is small enough to allow direct observation of stars in the galactic nucleus; the extinction at 2.2 $\mu$m is 2.7 magnitudes,[2] corresponding to 27 magnitudes of visual extinction.
2. The dust near the Galactic center absorbs the power radiated at ultraviolet and optical wavelengths;[3] most of the power will be absorbed in a region in which the absorption optical depth at visible wavelengths is of order unity. This power is radiated away by the dust at infrared wavelengths,[4] and the infrared emission is optically thin.

In this paper analysis of the thermal emission from the dust will be emphasized. In this way the total bolometric luminosity of the Galactic center and the distribution of emitting dust are deduced, and constraints are provided on the nature of the source of this luminosity.

### HEATING OF DUST BY HII REGIONS

Figure 1 shows the measured infrared energy distribution of the central 30" (1.4 parsecs) of the Galaxy (SGR A) in comparison with several galactic HII regions.[5] G333.6-0.2 and K3-50 are excited by early type stars recently formed in massive molecular clouds; their energy distributions are typical of HII regions in such areas of star formation.[6] NGC 7027 is a planetary nebula with an energy distribution representative of that class.[7]

The energy distribution gives both the total luminosity of the source (from the area under the curve) and a representative temperature of the emitting grains (from the wavelength at which the energy distribution peaks). For the present purposes it is most important that the mean grain temperature in the planetary nebula is higher than in the HII regions excited by young stars. This is true despite the fact that the total luminosity of the planetary nebula is less. This comparison illustrates the relative contributions to the heating of the dust by two important mechanisms, namely (1) resonantly trapped Lyman $\alpha$ (L$\alpha$) line radiation in the plasma and (2) direct heating by starlight, as follows.

Consider a radiatively ionised HII region where the dust optical depth at visible wavelengths inside the plasma is very much less than unity. Each Lyman continuum photon from the central star ionises a hydrogen atom and recombination eventually produces a L$\alpha$ photon, which, being a ground state transition, is resonantly trapped. This ensures that the L$\alpha$ radiation is absorbed by the small amount of dust which is present. Viewed from the perspective of a dust grain surrounded by plasma the L$\alpha$ radiation field is intense, and grains heated this way become relatively hot.

In a planetary nebula the central star is hot enough that much of the power of the central star is radiated in the Lyman continuum; heating by L$\alpha$ dominates and the grains appear hot.

The situation is rather different in the HII regions powered by young stars; in this case much of the total power of the stars is radiated beyond the Lyman limit. This radiation naturally proceeds to cross the ionisation fronts. Typically there is sufficient material immediately outside the ionised region left over from the star formation process to ensure that this flux of ultraviolet and visible radiation is absorbed by dust in a neutral shell (in which the dust optical depth is about unity). This large amount of power is radiated by the dust in the infrared; grains beyond the ionisation fronts heated in this way, directly by stellar photons, are generally found to be rather cooler than those inside the plasma.

Two further points regarding L$\alpha$ heating are worthy of note. First, the L$\alpha$ radiation field is produced by recombinations, and so its local intensity depends on the square of the electron density; density enhancements in the plasma ("clumps") are therefore expected to appear hot and bright in the infrared. Second, and consequently, the observation of infrared emission from hot grains in any particular clump of plasma does not demand the presence of an <u>embedded</u> star to ionise the clump; the clump may equally well be <u>ionised</u> by an exterior source. The position of the ionising source relative to the plasma clump cannot be inferred when L$\alpha$ heating is dominant.

## LUMINOSITY OF THE GALACTIC CENTER

The infrared energy distribution of the central 30" of the Galaxy, labelled SGR A in Figure 1, peaks at a wavelength of 30μm; the directly measured luminosity of this central region is $2 \times 10^6 L_\odot$. The similarity of this energy distribution to that of NGC 7027 is striking; this suggests immediately that heating by Lα radiation within the central HII region[10] (described by Lo) is an important energy input to the grains.

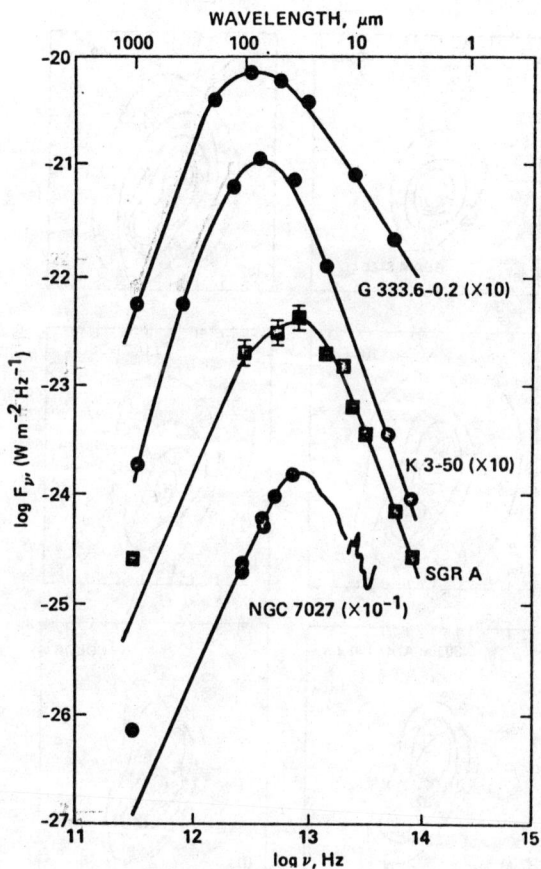

Figure 1. Energy distributions of galactic sources.

Figure 2a,b,c shows maps of the Galactic center with 30" resolution at wavelengths of 30μm, 50μm and 100μm.[5] The position of the Galactic center is indicated by a cross in each map. At 30μm the source is compact and centered on the Galactic nucleus, but at longer infrared wavelengths the source takes on a double-lobed appearance, with the lobes extended along the plane of the Galaxy. <u>The position of the Galactic center lies between the far infrared lobes.</u> The surface brightness distributions at 30μm and 100μm are compared directly in Figure 2e, and are clearly very different.

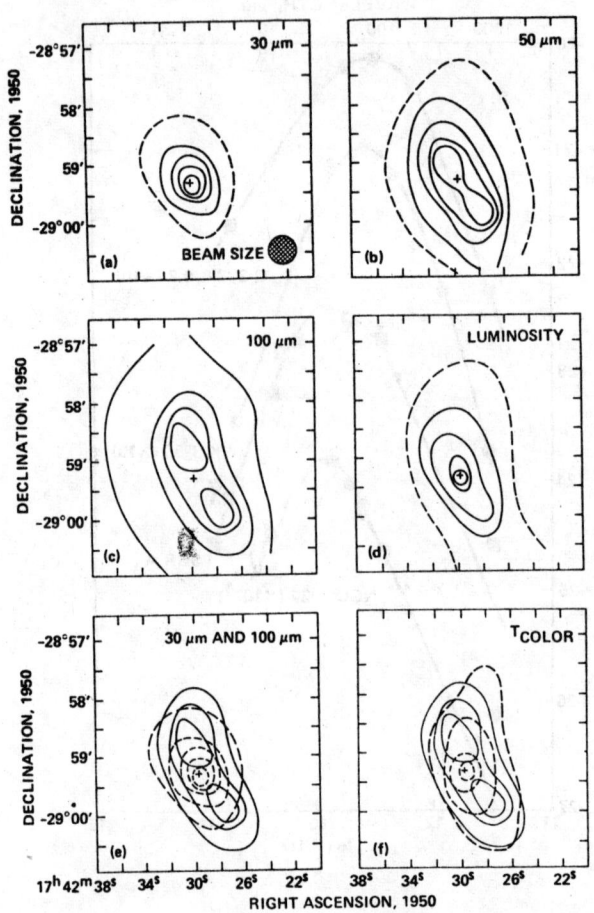

Figure 2. Maps of the Galactic Center

As described above, the measured energy distribution at each spatial position can be used to deduce both a luminosity and a temperature for the radiating grains. Figure 2d shows the distribution of luminosity, and Figure 2f shows a map of color temperature (obtained by fitting a blackbody curve through the 50μm and 100μm data) superposed on the 100μm surface brightness.

The peaks in color temperature and far infrared luminosity seen at the Galactic center suggest that the dust is heated primarily by a centrally concentrated source of luminosity. This idea is supported by the fact that the lobes seen at 100μm on either side of the Galactic center do not coincide with peaks in either temperature or luminosity, and so do not appear to contain substantial embedded internal energy sources. Rather, because a single grain radiates more power at every wavelength as its temperature is increased, and because the low optical depth of emission ensures that all radiating grains are observed, it must be true that there is a local minimum in the dust density in the central parsec of the Galaxy. The double-lobed appearance of the source at longer wavelengths arises naturally if the dust density is highest in the galactic plane. Non-ionising radiation streaming out of the central parsec (where the dust density is evidently too low to absorb it) enters a region of higher dust density which is shaped like a ring lying in the galactic plane. Seen in projection, the emission from this ring causes the double lobes.

The total luminosity of the sources of optical and ultraviolet radiation within the central parsec of the Galaxy can now be estimated. The total luminosity within the lowest solid contour of Figure 2d is $5 \times 10^6 L_\odot$. Under the assumption that this power is radiated from a ring of dust which subtends $2\pi$ sr as seen from the Galactic center the central luminosity is $10^7 L_\odot$. An upper limit to the central luminosity can be established from the observed color temperature of the lobes via a scaling argument from observations of other sources in which the grains are also directly heated.[5] From these considerations the central luminosity is found to lie in the range $1-3 \times 10^7 L_\odot$.

## ORIGIN OF THE LUMINOSITY

The far infrared observations allow a description of the Galactic center based simply on the properties of HII regions given above. In the central parsec the gas is ionised and the few grains present are heated chiefly by trapped $L_\alpha$. The distribution of plasma in the central parsec is clumpy[11] (as discussed by Lacy). The far infrared results show very clearly that the clumps cannot be embedded in dense neutral material, as are compact HII regions in areas of star formation in the spiral arms.[6,12] On the contrary the plasma clumps are density enhancements within a largely empty volume. Because the plasma clumps are not surrounded by dense neutral shells the possibility exists that each clump need not contain a source of ionisation. This is demonstrated clearly by

the fact that luminosity escapes the central parsec to heat the far
infrared lobes. In fact these lobes are produced by direct heating
beyond the ionisation fronts of Sgr A, as was described in general
terms earlier; their emission represents that power radiated in the
central parsec beyond the Lyman limit.

In contrast to the situation in the Galactic center, regions
of star formation in the spiral arms contain a great deal of
neutral material in the associated molecular clouds and the HII
regions are often embedded in a dense neutral shell. Then the
contributions from grains inside and outside the ionisation fronts
are not spatially resolvable, and energy distributions like those
of G333.6-0.2 and K3-50 (given in Figure 1) result. This
comparison further illustrates that the Galactic center is a region
of relatively low dust density. In particular, the possibility of
star formation in the Galactic center[3] is difficult to reconcile
with the conspicuous absence of high density material which so
characterises star formation elsewhere in the Galaxy.

The directly measured luminosity of the central parsec,
attributable chiefly to $L\alpha$ heating, is $2 \times 10^6 L_\odot$; the total
luminosity is $1-3 \times 10^7 L_\odot$. Therefore only about 10% of the total
radiative power of the central parsec is emitted in the Lyman
continuum. This implies that the energy distribution in the
radiation field in the central parsec is similar to that of a late
O or early B type star, which is consistent with the
spectroscopically determined limit of $T \sim 35000K$ on the effective
temperature of the ionising source.[11]

## LOCATION OF THE IONISING SOURCE

The picture presented here, in which (1) the central parsec is
transparent to ultraviolet radiation, (2) the plasma consists of
clumps in a largely empty region, and (3) the source which ionises
the gas heats the grains, leads us to expect that high spatial
resolution infrared maps of the central parsec at wavelengths
around $30\mu m$ (the peak of the energy distribution) should look
similar to radio maps of the free-free emission. This is the case;
Figure 3 (solid contours) is a map of the Galactic center at $34\mu m$[13]
with a spatial resolution of 8". The distribution of $34\mu m$ surface
brightness is similar to that seen in the VLA map of the free-free
emission[10] presented by Lo. Higher resolution infrared maps[3,14] at
$10\mu m$ and $20\mu m$ confirm this effect in more detail.[10]

Despite the obvious importance of $L\alpha$ heating in the Galactic
center, the large amount of power in non-ionising radiation flowing
through the central parsec raises the interesting possibility that
direct heating of grains will also have observable consequences in
this inner region. In this case temperature maps at wavelengths
around $30\mu m$ may give information on the location of the source or
sources of luminosity.

Figure 3 (dashed contours) shows the measured ratio of $20\mu m$ to
$34\mu m$ flux in the Galactic center;[13] this quantity is a measure of

the temperature of the emitting grains. The temperature contours are roughly circularly symmetric, and do not peak on the position of highest infrared surface brightness. Rather the temperature peak coincides with the position of the non-thermal radio source[10] and its probable 2.2μm infrared counterpart, IRS 16, which is believed to be the position of the nucleus.[14] This result suggests that the luminosity of the central parsec of the Galaxy originates from a source or distribution of sources concentrated at the very center of the Galaxy.

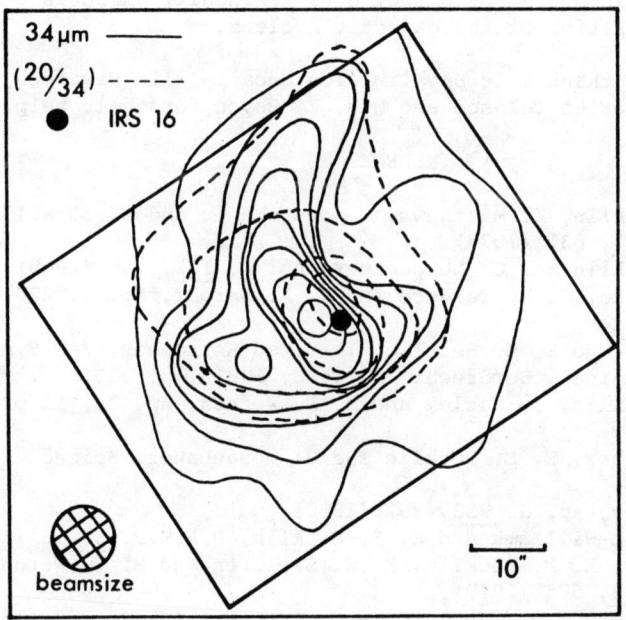

Figure 3. Grain color temperature in the Galactic Center

In summary, analysis of the thermal emission from the dust in the Galactic center provides information on the total bolometric luminosity, and the density, temperature and distribution of the dust itself. More detailed analysis of the conditions in the Galactic center can be usefully constrained by these results, summarised below.

## CONCLUSIONS

1. The total bolometric luminosity of the Galactic center is in the range $1-3 \times 10^7 L_\odot$.
2. About 10% of this luminosity is radiated in the Lyman continuum.
3. The dust density in the central parsec is very low; the visual extinction across the central parsec is much less than one magnitude.
4. The dust density increases with distance from the center in the central few parsecs, and is highest in the galactic plane.
5. The dust is heated by a source of luminosity which is centered on the position of the galactic nucleus.

I want to thank Eric Becklin, Tom Geballe, Tim Hawarden, Gerry Neugebauer, Charles Telesco and Mrs. Y. Boyce for their help.

## REFERENCES

1. E. E. Becklin, K. Matthews, G. Neugebauer and S. P. Willner, Ap. J. 220, 831 (1978).
2. E. E. Becklin and G. Neugebauer, Ap. J. 151, 145 (1968).
3. G. H. Rieke, C. M. Telesco and D. A. Harper, Ap. J. 220, 556 (1978).
4. I. Gatley and E. E. Becklin, Infrared Astronomy, IAU Symposium No. 96 (Reidel, Dordrecht, Holland, 1981), p. 281.
5. E. E. Becklin, I. Gatley and M. W. Werner, Ap. J., in press, 1982.
6. M. W. Werner, E. E. Becklin and G. Neugebauer, Science 197, 723.
7. H. Moseley, Ap. J. 238, 892 (1980).
8. C. G. Wynn-Williams and E. E. Becklin, P.A.S.P. 86, 5 (1974).
9. I. Gatley, E. E. Becklin, K. W. Sellgren and M. W. Werner, Ap. J. 233, 575 (1979).
10. R. L. Brown, K. J. Johnston and K. Y. Lo, Ap. J. 250, 155 (1981).
11. J. H. Lacy, C. H. Townes, T. R. Geballe, and D. J. Hollenbach, Ap. J. 241, 132 (1980).
12. H. A. Thronson, Jr. and D. A. Harper, Ap. J. 230, 133 (1979).
13. I. Gatley, E. E. Becklin and C. M. Telesco, unpublished.
14. E. E. Becklin and G. Neugebauer, Ap. J. (Letters) 200, L71.

NEW FAR INFRARED OBSERVATIONS OF THE CENTRAL 30' OF THE GALAXY

W. A. Dent
University of Massachusetts, Amherst, Ma. 01003

M. W. Werner
NASA Ames Research Center, Moffett Field, Ca. 94035
California Institute of Technology, Pasadena, Ca. 91125

I. Gatley
UKIRT, Hilo, Hi. 96720
California Institute of Technology, Pasadena, Ca. 91125

E. E. Becklin
University of Hawaii, Honolulu, Hi. 96822

R. H. Hildebrand
University of Chicago, Chicago, Il. 60637

J. Keene
S. E. Whitcomb
University of Chicago, Chicago, Il. 60637
California Institute of Technology, Pasadena, Ca. 91125

## ABSTRACT

A 45' x 30' region around the galactic center was mapped with 1' resolution at 55 μm and 125 μm using the Kuiper Airborne Observatory. Peaks in temperature of the dust are correlated with centimeter wavelength thermal continuum sources. The distribution of the column density of dust shows minima at the galactic center (Sgr A) and at the position of an HII region complex (G.07+.04) 10' to the North.

## INTRODUCTION

The bulk of the observed luminosity of the central 30' of the Galaxy is known to be at far infrared wavelengths between 30 and 200 microns. This infrared radiation is thermal emission from dust grains heated by a variety of luminosity sources. Previous far infrared observations of this region of the Galaxy have identified emission associated with background late type stars, HII regions, molecular clouds, and diffusely distributed dust. In this complex environment far infrared observations at several wavelengths are valuable because of the additional diagnostic information which can be extracted from the distribution of dust temperature and optical depth.

The present paper presents the results of observations with a 1' resolution at 55 μm and 125 μm of the central 30' of the

Galaxy. These new observations extend the spatial coverage, with greater sensitivity, of the previous far infrared maps of Gatley et al (1977, hereafter GBWW) and complement the larger spatial scale broadband results recently presented by Odenwald et al (1980).

## OBSERVATIONS

A 45' x 30' rectangular region around the galactic center was mapped in 2 wavelength bands using the 91cm infrared telescope of NASA's Kuiper Airborne Observatory. The observations were made in May 1979. The half power band widths were 25 μm and 205 μm centered at effective wavelengths of 55 μm and 125 μm respectively. The flux density calibration was made relative to the planet Uranus which was assumed to have brightness temperatures of $61°K$ and $59°K$ at 55 μm and 125 μm.

The instrumentation used for the observations was the University of Chicago six-channel bolometer system (Whitcomb et al, 1981), with three detectors in each wavelength band. Three separate but nearly adjacent 1' fields of view on the sky were observed simultaneously, and a dichroic beam splitter made it possible to observe each field of view simultaneously at both wavelengths. To minimize the effects of extended emission, the chopper throw was set to 10', and the data were taken by scanning parallel to the chopper direction, which lay roughly along lines of constant declination. Twenty such scans were carried out covering somewhat more than 20' along the galactic plane; the three fields of view projected onto the sky in such a way that the final map was fully sampled. The 45' long scans were deconvolved by assuming a constant baseline except in a few cases where drifts necessitated use of a linearly sloping baseline. The RMS noise fluctuations at each spatial point in the final deconvolved scans were 100 Jy per 1' beam at 125 μm and 300 Jy per 1' beam at 55 μm. To produce the flux distribution shown in Figure 1, the 55 μm data has been smoothed to an effective resolution of ~2', except in the bright central region where the original 1' resolution was retained.

## RESULTS

A. Flux Distribution

The 55 μm and 125 μm maps are shown in Figures 1a and 1b respectively. The principal features of the flux distributions are as follows:

1) At both wavelengths the flux peaks sharply in the vicinity of the galactic center and shows considerable extension along the galactic plane.

2) The peak 125 μm emission is seen not at the position of the galactic center (defined by Sgr A West and the infrared cluster) but 30" to the south with the central contours

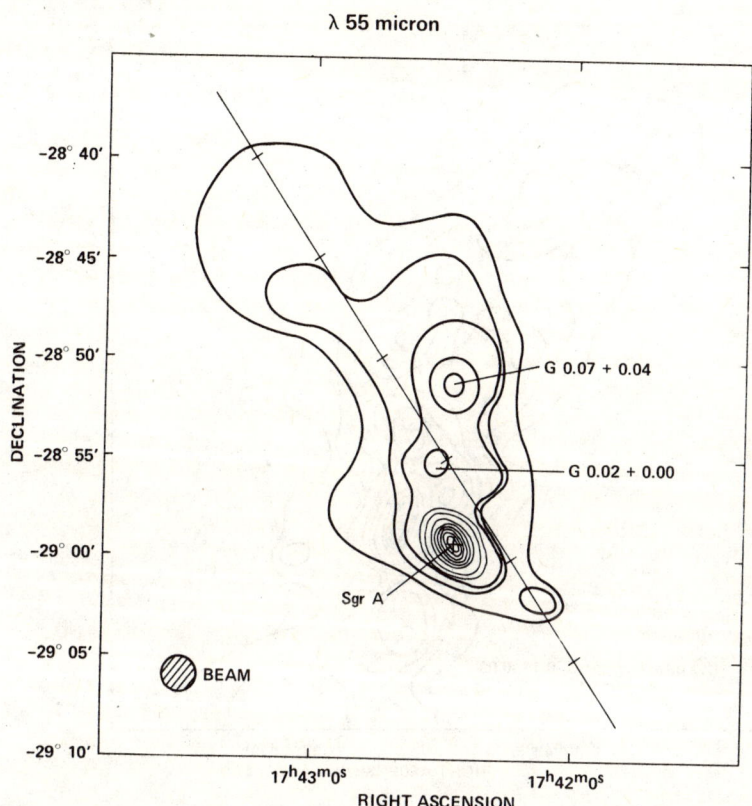

Fig. 1a. The 55 μm emission map of the galactic center region. The contours are 1000, 1500, 2000, 2500, 3000, 4000, 5000, 6000, 7000, 8000, and 9000 Janskys per 1' beam (1000 Jy per 1' beam corresponds to $1.5 \times 10^{-14}$ $Wm^{-2}$ $Hz^{-1}$ $sr^{-1}$). The cross indicates the position of the galactic center.

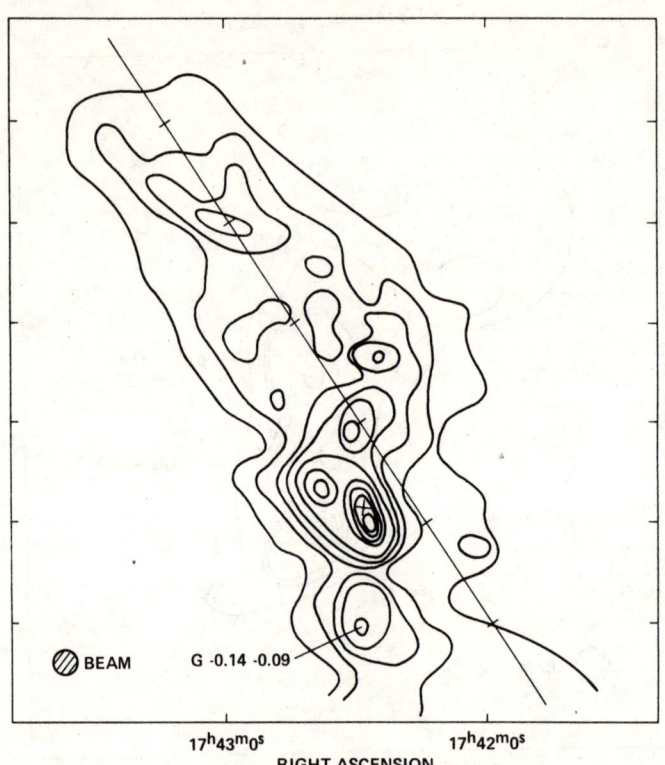

Fig. 1b. The 125 μm emission map. The contours are 1000, 1500, 2000, 2500, 3000, 3500, 4000, 4500, and 5000 Janskys per beam solid angle (1000 Jy per 1' beam corresponds to $1.5 \times 10^{-14}$ $Wm^{-2}$ $Hz^{-1}$ $sr^{-1}$). The cross indicates the position of the galactic center.

strongly elongated along the galactic plane. These
results are consistent with the recent high resolution
map of Becklin, Gatley and Werner (1982) which show that
at 100 μm the structure of Sgr A consists of two peaks
aligned parallel to the galactic plane and situated ±30"
on either side of the galactic center. As discussed
later, the lower 125 μm flux density at the galactic
center reflects a lower dust column density.
3) There is a marked asymmetry about Sgr A along the galactic
plane. The source at l = .14, b = -.09 some 5' south of
Sgr A is strong at 125 μm but not seen at 55 μm. It is
noteworthy that the 540 μm submillimeter emission peaks
at the same location (Hildebrand et al, 1978).
4) Only a few of the features at 125 μm correlate well
with the high resolution centimeter wavelength maps of
Pauls et al (1976) and Kapitzky and Dent (1974). These
are the galactic center (Sgr A West), the HII region
G.07+.04, and the 125 μm source G.02+.00 which coincides
with an emission minimum at centimeter wavelengths.
There is, however, a much stronger general correlation
between the 55 μm and the centimeter wavelength features.
There is no evidence at either 55 μm or at 125 μm for the
centimeter wavelength sources Sgr A East and G.16-.15,
both of which are non-thermal and hence probably supernovae
remnants.

B. Dust Temperature and Dust Opacity

The variation in dust temperature across the region was
computed from the 55 μm : 125 μm flux density ratio assuming an
emission efficiency of the dust which varies as $\lambda^{-1}$. This
distribution is shown in Figure 2a along with some of the 55 μm
emission contours (dashed lines) and the locations of peaks
in thermal bremsstrahlung emission (plus signs) mapped at 10.6 GHz
by Pauls et al (1976). The temperature peaks sharply at the
position of Sgr A and shows subsidiary peaks associated with
several features of the 55 μm flux distribution, most notably
with the HII region complex G.07+.04 about 10' north of Sgr A.
Apart from these peaks, the temperature is very uniform at ~40K
within the outer contours of the 55 μm map. There is also
a general association between the centimeter wavelength sources
and regions of higher dust temperature implying a common heating
source for both the dust and gas at these locations. Since
untraviolet photons are required to ionize the gas, the common
energy sources are probably early type stars.

The distribution of the line of sight column density of
dust was computed at each location by solving the equation of
radiative transfer for the opacity. The overall distribution
of dust bears a close similarity to the lower resolution
Hildebrand et al (1978) map of the 540 μm emission near the
galactic center and supports the existence of the
features in the present opacity map.

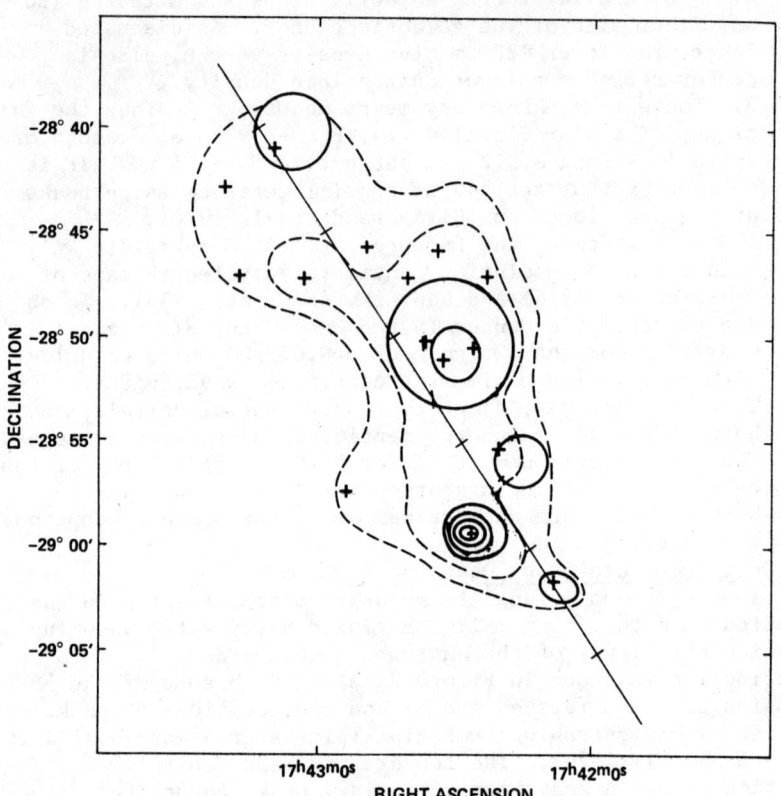

Fig. 2a. The temperature distribution of the emitting dust (solid lines). The outer 55 μm contours (dashed) are shown for reference. The plus signs indicate the location of centimeter wavelength emission peaks (Pauls et al, 1976). The contours start at 45°K and increase in steps of 5°K. The ambient temperature within the region of emission is 40°K.

## DUST OPACITY AND AMMONIA EMISSION

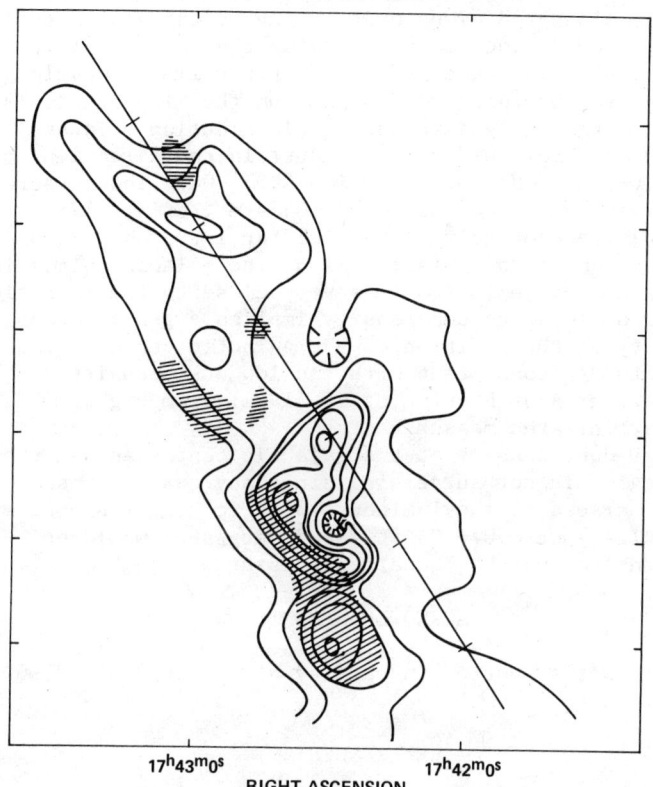

Fig. 2b. The distribution of column density of dust expressed as an opacity at 125 μm. The contours are $\tau$ = .012, .018, .024, .030, .036, .042, and .048 and were computed using the temperature distribution in Figure 2a.

The shaded areas in the opacity map show the spatial distribution of emission from the ammonia molecule (Gusten, Walmsley and Pauls, 1982). The distribution of ammonia as well as those of formaldehyde (Gusten and Downes, 1980) and HCN (Fukui et al, 1977) are highly correlated with the distribution of dust near the galactic center. This is especially evident in the elongated cloud east of the center and in the dense cloud G-.14-.09 located 5' south of the center. This latter dust cloud is not seen at 55 μm or centimeter wavelengths and appears as an absorbing void at 2.2 μm (Becklin et al, 1978).

The most striking feature in the distribution of dust opacity is the low column density of dust in the directions of the galactic center and the HII region G.07+.04. The measured opacities toward the center and G.07+.04 are $\tau_{125}$ = .017 and $\tau_{125}$ = .011 respectively. The high far infrared luminosity and the elongation of the emission along the galactic plane imply that the dust seen in emission lies very close to the galactic center; thus the lower column density implies a genuine drop in dust density at the position of the galactic center. The present results are consistent with the low dust density determined for Sgr A by Becklin et al (1982) from higher resolution far infrared measurements.

The lower dust density at the galactic center and in the nearby HII region is not surprising since any dust grains within a few parsecs of the luminous galactic center source (L = $10^{41}$ erg/sec) or G.07+.04 (L = $10^{40}$ erg/sec) would be destroyed or driven out by radiation pressure.

## ACKNOWLEDGEMENT

This work was supported in part by NASA grants NGR 05-002-281 and NSG-2057.

## REFERENCES

1. E. E. Becklin, I. Gatley, and M. W. Werner, Ap. J. (in press) (1982).
2. E. E. Becklin and G. Neugebauer, Publ. Astron. Soc. Pacific 90, 657 (1978).
3. Y. Fukui, T. Iguchi, N. Kaifu, Y. Chikada, M. Marimoto, K. Nagane, K. Miyazawa, and T. Miyaji, Pub. Astron. Soc. Japan 29, 643 (1977).
4. I. Gatley, E. E. Becklin, M. W. Werner and C. G. Wynn-Williams, Ap. J. 216, 277 (1977).
5. R. Gusten and D. Downes, Astron. Astrophys. 87, 6 (1980).
6. R. Gusten, C. M. Walmsley and T. Pauls, Astron. Astrophys. (in press) (1982).
7. R. H. Hildebrand, S. E. Whitcomb, R. Winston, R. F. Stiening, D. A. Harper and S. H. Moseley, Ap. J. 219, L101 (1978).
8. J. E. Kapitzky and W. A. Dent, Ap. J. 188, 27 (1974).

9. S. Odenwald, G. G. Fazio and E. L. Wright, Bull. Amer. Astron. Soc. $\underline{12}$, 860 (1980).
10. T. Pauls, D. Downes, P. G. Mezger and E. Churchwell, Astr. Ap. $\underline{46}$, 407 (1976).
11. S. E. Whitcomb, I. Gatley, R. H. Hildebrand, J. Keene, K. Sellgren and M. W. Werner, Ap. J. $\underline{246}$, 416 (1981).

## LARGE BEAM OBSERVATIONS OF THE GALACTIC CENTER
## AT 150, 200, AND 300 μm

M. T. Stier[1], E. Dwek[1], R. F. Silverberg, M. G. Hauser,
L. Cheung[2], T. Kelsall, and D. Y. Gezari
NASA/Goddard Space Flight Center, Greenbelt, MD 20771

### ABSTRACT

We present 10' resolution observations of a 2 square degree portion of the galactic plane around the galactic center at 150, 200, and 300 μm, obtained as part of a galactic plane survey. In the galactic center region the mean dust temperature is ∼30 K. Both the temperature and the dust-to-gas mass ratio in the galactic center region are typical of a large portion of the galactic plane. Three main emission peaks, Sgr A, B, and C, are detected; these peaks appear as dust column density enhancements rather than temperature enhancements. The dust responsible for the far-infrared emission may only be a fraction of the dust responsible for the visual extinction toward the galactic center.

### INTRODUCTION

Interest in the infrared emission from the galactic center is of long standing and has been studied by a variety of investigators[1] (and references therein). We report here on 10' resolution multi-band observations of the galactic center region that allow the determination of the temperature and column density of the dust responsible for the majority of the observed luminosity. Our extensive galactic plane coverage of both discrete sources and diffuse emission allows a direct comparison of the conditions near the galactic center with other portions of the galactic plane.

### OBSERVATIONS

The far-infrared (far-IR) observations of the galactic center were made on 18 August 1980 with the NASA Goddard Space Flight Center 1.2 m inertially-stabilized balloon-borne telescope[2] and a 3-band far-IR photometer containing long-wavelength-pass filters with cut-on wavelengths of 110, 150, and 260 μm. The beam size for each channel was 10'x 10'. Previous observations[3] of the galactic plane were made on 16 November 1979. Calibration of the galactic center data is based on scans of M17. The M17 brightness distribution was determined during the 1979 balloon flight when both M17 and Jupiter (assumed to be a 125 K black body) were observed. The rms noise in the short, medium, and long wavelength maps (Fig. 1 a-c) is 650, 760, and 1200 Jy/beam respectively.

[1] National Research Council Resident Research Associate
[2] Present address: Gulf Oil Corp., Pittsburgh, PA

## DISCUSSION

In Figure 1 10' resolution maps of the galactic center region at effective wavelengths of 150 μm (1a), 200 μm (1b), and 300 μm (1c) are shown. At all three wavelengths the spatial distribution of the far-IR emission is very similar to that of the integrated $^{13}$CO emission[4]. Comparison of the far-IR maps with the 11' resolution 5 GHz radio continuum maps[5] indicates that the far-IR and molecular emission have somewhat better correlation than the far-IR and radio continuum emission. The differences between the far-IR and the radio continuum emission may result from the substantial non-thermal emission at 5 GHz in the galactic center region[6].

Table I shows observed parameters of the three far-IR sources. The uncertainty in source positions is about 0.1°. Scaling the flux density at 150 μm to account for the different beam size and wavelength yields good agreement with previous observations[7] of Sgr A with a 5' beam at 105 μm.

Table II lists several quantities of astrophysical interest derived from our observations. Assuming a dust spectral emissivity index of 1.6 to scale the 150 μm flux for emission outside the passband yields a bolometric luminosity for the entire 1° x 2° region in Fig. 1 of $4.4 \times 10^8$ $L_\odot$. This is somewhat larger than the value of $\sim 3 \times 10^8$ $L_\odot$ measured previously in a 75-125 μm passband[8].

The temperature of the emitting dust is determined from the ratio of the 150 μm intensity to the 300 μm intensity assuming a spectral emissivity index of 1.6. In the galactic center region the mean temperature of the dust is $\sim 30$ K. This is similar to the average dust temperature of 29 K (5 K sample standard deviation) found for several positions on the galactic plane between 0° and 40° longitude. This dust temperature is comparable to but lower than the 35-45 K temperatures reported for the energetic galaxies NGC 1068, NGC 253, and M82[9].

The mass column density of dust is proportional to the optical depth and inversely proportional to $\kappa(300\mu m)$, the mass absorption coefficient ($cm^2 g^{-1}$) of dust at 300 μm. The <u>total</u> dust column densities, $M_d$, listed in Table II were calculated using a silicate-to-carbon dust mass ratio of 5:2[10] and assuming that silicates dominate the far-IR emission with $\kappa(300\mu m)=43$[11]. The value of $\kappa(300\mu m)$ is, however, very uncertain and may be as low as $\sim 10$[9]. If the emissivity in the far-IR is dominated by amorphous carbon grains with $\kappa(300\mu m)=26$[12], then the resulting values of $M_d$ in Table II would have to be increased by a factor of about 4.

The large beam far-IR observations, together with other infrared observations of the galactic center reveal the existence of several dust components toward that region. The first component consists of cool ($\sim 30$K) dust that gives rise to the far-IR emission detected in our survey bands. As the temperatures of Sgr A, B, and

C are similar to the mean temperature in the galactic center region, and as the 300 μm flux (which is not strongly temperature dependent) shows significant enhancement at these positions, the emission peaks must be largely due to density enhancements along the line of sight. The dust temperatures at the peaks of Sgr A and Sgr C (Table II) are about one-half of the temperatures measured with a ∼1' beam[13]. Thus, the small beam observations suggest a second, warm (∼70 K) dust component radiating from a small region around these sources. A comparison of the infrared luminosities of the two dust components implies that when averaged over a 10' region the emission is dominated by the cooler dust around the sources. The large beam observations of Sgr B yield a dust temperature of 26 K; this is only slightly cooler than the 32 K temperature measured[13] with a small beam. This does not rule out the presence of a warm dust component in Sgr B2 since the source is optically thick at 100μm[14]. The 10' beam observations of Sgr B[14] indicate that the radial decrease in temperature observed[14] with 0.5' to 1.5' beams continues to significantly larger radii.

The existence of a third cold dust component may be inferred by comparing the dust-to-gas mass ratio derived from the 300 μm observations with[15] the extinction in the 9.7 μm silicate absorption feature observed[15] toward the galactic center. Multiplying $A_v$ = 30 mag toward the galactic center[16] by a factor of 2 to account for dust on the far side of the galactic center and taking[17,11] $\kappa(9.7\mu m)=3 \times 10^3$, the total dust-to-gas mass ratio is $2.9 \times 10^{-3}$. If the far-IR emission is dominated by silicate grains[11] with $\kappa(300\mu m)=43$ then our observations toward Sgr A suggest the presence of a third, cold (T<30 K) dust component that gives rise to the visual extinction towards the galactic center but does not significantly contribute to the 300μm emission. A comparison of the respective dust-to-gas mass ratios shows that this third cold component is about five times more abundant than the cool dust that is detected at 300 μm. To make this comparison we corrected for the presence of carbon grains intermixed with the silicates that give rise to the 9.7 μm absorption feature. However, if amorphous carbon grains dominate the far-IR emission[12] or $\kappa(300\mu m)=10$ for silicates, then the column density of dust that contributes to the visual extinction is comparable to the column density of dust that gives rise to the far-IR emission.

It has recently been suggested that the dust-to-gas mass ratio should exhibit a substantial increase toward the galactic center[18]. Using $^{13}$CO observations[19] at several longitudes along the galactic plane to derive gas column densities, the dust-to-gas mass ratio toward the galactic center is similar to the ratio along the galactic plane between 10° and 40° longitude. If the dust mass derived from the far-IR emission is a constant fraction of the total amount of dust then the dust-to-gas mass ratio has no systematic trend with longitude. On a smaller scale both the dust mass and the dust-to-gas mass ratio observed toward Sgr B are larger by a factor of ∽3 than that observed toward Sgr A or C. One possible explanation for the high observed ratio is that the $^{13}CO$ emission used to derive the gas column density toward Sgr B arises from saturated lines. Alternatively the results could reflect real differences between the sources in the dust-to-gas mass ratio. Observations of optically thin $C^{18}O$ emission from Sgr B2 would be useful in settling this question.

## ACKNOWLEDGMENTS

We would like to thank D. Walser for engineering support and L. Blitz and D. Jaffe for useful discussions. The National Scientific Balloon Facility was responsible for the launch, tracking, and recovery of the Goddard Submillimeter Survey telescope.

## REFERENCES

1. Gatley, I., and Becklin, E. E., Infrared Astronomy (ed. C. G. Wynn-Williams and D. P. Cruikshank, I.A.U., 1981), p. 281.
2. Silverberg, R. F., Hauser, M. G., Mather, J. C., Gezari, D. Y., Kelsall, T., and Cheung, L., Proc. S.P.I.E., 172, 149 (1979).
3. Cheung, L., NASA/GSFC TM-82056 (1980).
4. Heiligman, G. M., The Phases of the Interstellar Medium (ed. J. M. Dickey, National Radio Astronomy Observatory, Green Bank, 1981), p. 93.
5. Altenhoff, W. J., Downes, D., Goad, L., Maxwell, A., and Rinehart, R., Astr. Ap. Suppl., 1, 319 (1970).
6. Mezger, P. G., and Pauls, T., The Large-Scale Characteristics of the Galaxy (ed. W. B. Burton, I.A.U., 1979), p. 357.
7. Harper, D. A., and Low, F. J., Ap. J. (Letters), 182, L89 (1973).
8. Hoffmann, W. F., Frederick, C. L., and Emery, R. J., Ap. J. (Letters), 164, 123 (1971).
9. Telesco, C. M., and Harper, D. A., Ap. J., 235, 392 (1980).
10. Mathis, J. S., Rumpl, W., and Nordsieck, K. H., Ap. J., 217, 425 (1977).
11. Mezger, P. G., Mathis, J., and Panagia, N., preprint (1981).
12. Draine, B. T., Ap. J., 245, 880 (1981).
13. Gatley, I., Becklin, E. E., Werner, M. W., and Harper, D. A., Ap. J., 220, 822 (1978).
14. Harvey, P. M., Campbell, M. F., Hoffmann, W. F., Ap. J., 211, 786 (1977).
15. Becklin, E. E., Mathews, K., Neugebauer, G., and Willner, S. P., Ap. J., 220, 831 (1978).
16. Becklin, E. E., and Neugebauer, G., Ap. J. (Letters), 157, L31 (1968).
17. Day, K. L., Ap. J., 234, 158 (1979).
18. Blitz, L., and Shu, F., Ap. J., 238, 148 (1980).
19. Solomon, P. M., Scoville, N. Z., and Sanders, D. B., Ap. J. (Letters), 232, L89 (1979).

Table I  Observed Properties of Galactic Center Sources

| Source | $\lambda_{eff}(\mu m)$ | $\ell^{II}$ | $b^{II}$ | Peak[A] Flux Density(Jy) | $\tau$[B] | FWHM[C] $\Delta\ell^{II}$ | $\Delta b^{II}$ |
|---|---|---|---|---|---|---|---|
| Sgr A | 150 | 359.°9 | -0.°2 | $2.3 \times 10^5$ | $4 \times 10^{-2}$ | 0.°4 | 0.°25 |
|       | 200 | 359.9  | -0.1  | $1.3 \times 10^5$ | $2 \times 10^{-2}$ |       |       |
|       | 300 | 359.9  | -0.1  | $4.8 \times 10^4$ | $1 \times 10^{-2}$ |       |       |
| Sgr B | 150 | 0.5    | -0.1  | $3.1 \times 10^5$ | $1 \times 10^{-1}$ | 0.25  | 0.25  |
|       | 200 | 0.5    | -0.2  | $2.0 \times 10^5$ | $7 \times 10^{-2}$ |       |       |
|       | 310 | 0.5    | -0.2  | $9.5 \times 10^4$ | $4 \times 10^{-2}$ |       |       |
| Sgr C | 150 | 359.5  | -0.3  | $1.7 \times 10^5$ | $3 \times 10^{-2}$ | ~0.4  | 0.2   |
|       | 200 | 359.5  | -0.2  | $1.1 \times 10^5$ | $2 \times 10^{-2}$ |       |       |
|       | 300 | 359.4  | -0.3  | $3.6 \times 10^4$ | $1 \times 10^{-2}$ |       |       |

Notes to Table I:
(A) Measured into a 10'x10' beam.
(B) Optical depth at effective wavelength assuming a spectral emissivity index of 1.6
(C) Non-deconvolved full-width at half-maximum after diffuse emission has been subtracted.

Table II  Derived Properties of Galactic Center Sources

| Source | $T(K)$[A] | $L(L_\odot)$[B] | $M_d(g\ cm^{-2})$[C] | $N_H(cm^{-2})$[D] | $M_d/M_g$[E] |
|---|---|---|---|---|---|
| Sgr A | 33 | $3.7 \times 10^7$ | $4.0 \times 10^{-4}$ | $3.0 \times 10^{23}$ | $6 \times 10^{-4}$ |
| Sgr B | 26 | $3.2 \times 10^7$ | $1.3 \times 10^{-3}$ | $3.3 \times 10^{23}$ | $2 \times 10^{-3}$ |
| Sgr C | 32 | $2.6 \times 10^7$ | $3.2 \times 10^{-4}$ | $2.6 \times 10^{23}$ | $5 \times 10^{-4}$ |

Notes to Table II:
(A) Derived from the ratio of 150 μm peak flux to 300 μm peak flux assuming a spectral emissivity index of 1.6.
(B) Bolometric luminosity (derived from flux into a 10'x10' beam centered on peak) for an assumed distance of 10 kpc.
(C) Total mass column density, adopting $\kappa(300\mu m)=43\ cm^2 g^{-1}$ and using the Mathis, Rumpl, and Nordsieck (1977) interstellar dust model.
(D) Derived from a mean $\int T_A^*(^{13}CO)dV$ (obtained by averaging the data of Heiligman (1981) over a 10'x10' region centered on the CO peaks). $N_H = 1.8 \times 10^{21} \int T_A^*(^{13}CO)dV\ cm^{-2}$
(E) Total dust-to-gas mass ratio derived from (C) and (D).

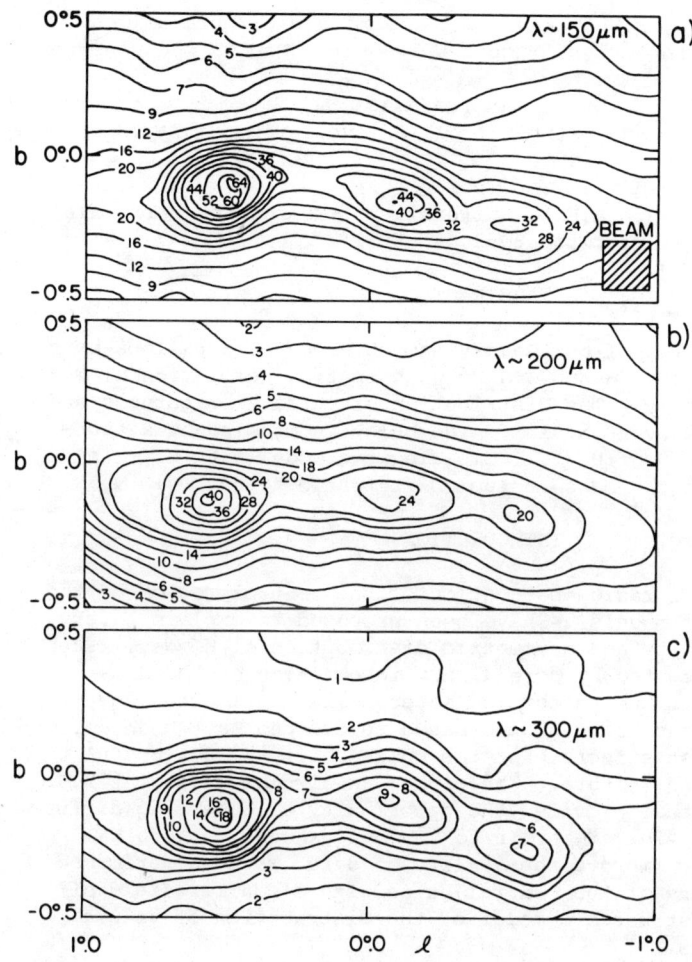

Fig. 1 Maps of the galactic center region at the effective wavelengths 150 μm (Fig. 1a), 200 μm (Fig. 1b), and 300 μm (Fig. 1c). The contour unit is $5.9 \times 10^{-18}$ W m$^{-2}$ Hz$^{-1}$ sr$^{-1}$, or $5 \times 10^3$ Jy from an equivalent 30K, n=1.6 point source. The beam size is 10'x10' at all three wavelengths.

# BALLOON OBSERVATION OF THE CENTRAL BULGE OF OUR GALAXY IN NEAR INFRARED RADIATION

T. Matsumoto, S. Hayakawa, H. Koizumi and H. Murakami
Department of Astrophysics, Nagoya University, Nagoya 464, Japan

K. Uyama and T. Yamagami
Institute of Space and Astronautical Science, Tokyo 153, Japan

J.A. Thomas
School of Physics (R.A.A.F.), University of Melbourne, Parkville,
Victoria 3052, Australia

## ABSTRACT

The central bulge of our Galaxy was observed by a balloon-borne infrared telescope at 2.4 and 3.4 μm with spatial resolution of 0.5° and 0.8°, respectively. The distribution of surface brightness and color are presented over $|l|<10°$. The bulge is modeled by a spheroid with axial ratio of 0.75. Comparing the model with HI rotation velocity, we obtain the mass to luminosity ratio to be 2.0±0.4.

## INTRODUCTION

On the basis of radio observation of the central region of the Galaxy, a number of models[1,2] have been proposed to explain a very high rotation velocity and asymmetric distribution of HI gas. Such models have to be tested by more direct information on the distribution of stars. Due to the low interstellar extinction, near infrared observations using ground-based telescopes have been able to reveal the innermost structure of the bulge[3], but are not useful to observe the extended feature. The use of an airglow window at 2.4 μm at balloon altitude provided the opportunity of observing diffuse galactic emission[4,5,6,7] in order to find an extended spheroidal bulge. In order to improve the quality of data, we have performed two color photometry of the central bulge with a smaller field of view, and have constructed a model of the distribution of volume emissivity and mass.

## OBSERVATION AND RESULTS

The observation was made with a balloon-borne infrared telescope at wavelength of 2.4 and 3.4 μm with bandwidth(FWHM) of 0.09 and 0.14 μm, respectively. The optical system was composed of a 15cm, F/1.0 lens system with 4 InSb detectors on the focal plane which provide the spatial resolution of 0.5° and 0.8° for 2.4 and 3.4 μm, respectively. The whole system was cooled by liquid nitrogen to reduce the thermal background noise, and output signals from the detectors were DC amplified without chopping.

The infrared telescope was launched on 2 and 13 April 1980 from the

Australian Balloon Launching Station at Mildura, Australia to raster scan the region of $|l|<15°$. Fig. 1 shows observed surface brightness in the form of a contour map at 2.4 µm with 0.5° field of view. The ratio of 2.4 to 3.4 µm flux values obtained with 0.8° field of view is shown in Fig. 2.

## INTERSTELLAR EXTINCTION

Since near infrared radiation in the bulge is undoubtedly of stellar origin and its color depends little on the spectral type in the observed wavelength region, the observed color primarily represents reddening.

In Fig. 1 fine structure is prominent mainly near the galactic plane, and the bright and dark regions are grossly identified with the blue and red regions in Fig. 2. This means that fine structure can be ascribed to the non-uniform distribution of dust in the disk. Our result can be verified by a survey of M-giants near the galactic plane[8].

Interstellar extinction to the galactic center obtained from the observed color is $A_{2.4}=1.8\pm0.3$ or $A_V=22\pm3$, and the average extinction over the region $|l|<1.0°$ is $A_V=15\pm4$ mag. The latitude distribution of extinction used for model calculation is obtained by adopting the gas to dust ratio of Dickman[9] and utilising the method

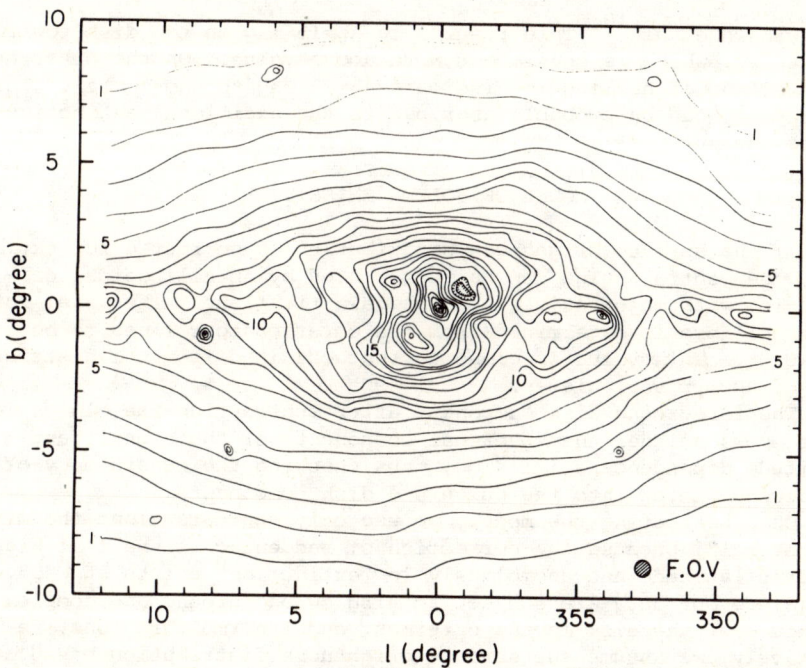

Fig. 1 Contour map of 2.4 µm surface brightness with 0.5° field of view. The lowest level corresponds to $1.0\times10^{-10}$ W cm$^{-2}$µm$^{-1}$sr$^{-1}$.

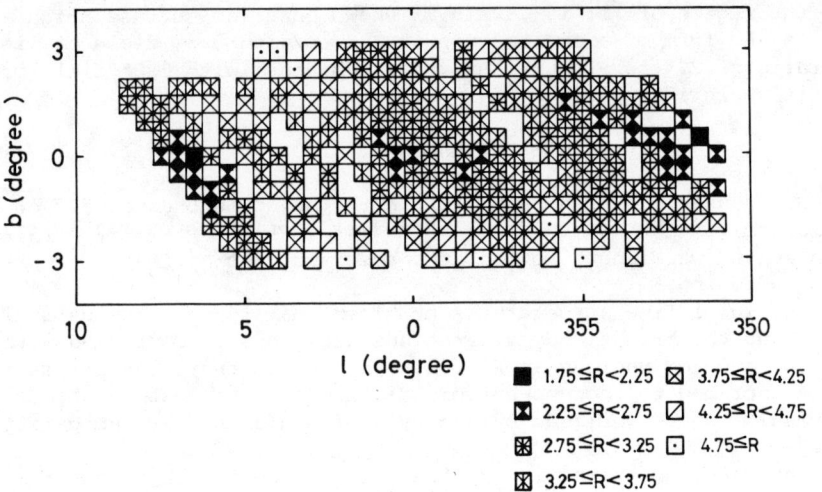

Fig. 2 Distribution of observed color. Color is represented by R, the ratio of observed surface brightness at 2.4 μm to that at 3.4 μm. The regions where 3.4 μm flux is not available are left blank.

of Hayakawa et al.[10] This results in $A_V$=15 mag in the disk towards the center, which is consistent with above value. On the other hand, observations of nucleus[3,11] indicate $A_V$=27 mag to the center. This discrepancy can be probably ascribed to the additional extinction in the nucleus.

## MODEL OF VOLUME EMISSIVITY

If the surface brightness distribution is corrected for extinction, the central bulge can be represented by an ellipsoidal distribution with major axis along the galactic plane. The axial ratio of the ellipse is determined from the outer contour lines to be 0.75+0.05. This value is close to that obtained from the distribution of RR Lyr variables[12].

The foreground disk component after subtracting the ellipsoidal bulge shows neither the spherical component nor the significant longitude dependence at $|l|<10°$, thus checking the consistency of the decomposition into the bulge and disk components.

On constructing the model, we use only the data along the minor axis at b>1°, because the correction of reddening is small at high galactic latitude and anomolously low extinction[8] exists at b<0°.

Since our data are subject to airglow background and complex structure of the disk in the outermost and innermost regions, respectively, we assume the surface brightness distribution $b^{-0.8}$ at b<1.0°[3] and $b^{-2}$ at b>5.0°[12,13]. By using a polynomial approximation and least square fitting, the following empirical formula for the volume emissivity at 2.4 μm, $\varepsilon_{2.4}$ is derived

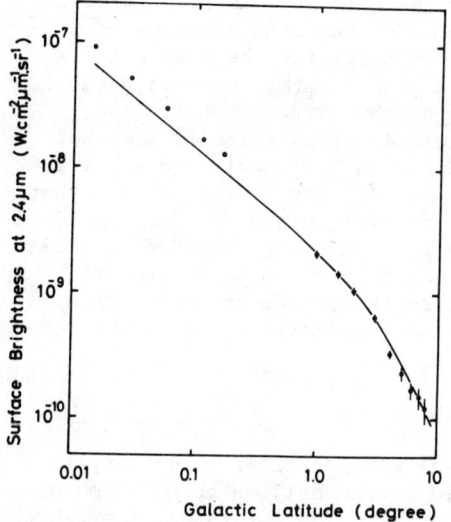

Fig. 3 Latitude distribution of the surface brightness corrected for extinction at 2.4 μm on l=0° plane. Solid and open circles indicate the data of ours and Becklin and Neugebauer[3], respectively. Straight line represents the model calculation.

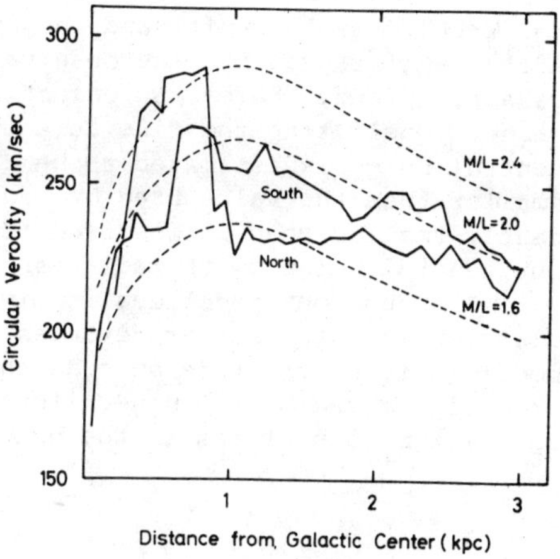

Fig. 4 The rotation velocity curve calculated for the model volume emissivity and given M/L in comparison with the observed one based on HI radial velocity by Sinha[14]. The same color as the central bulge of M31[15] is assumed.

$$a < 0.233 \quad \varepsilon_{2.4}(a) = 2.45 \times 10^{34}/a^{1.8} \text{ W (kpc)}^{-3} \mu m^{-1}$$

$$0.233 < a < 1.163 \quad \varepsilon_{2.4}(a) = -1.86 \times 10^{34} + 2.71 \times 10^{34}/a$$
$$+ 1.24 \times 10^{34}/a^2 + 1.46 \times 10^{32}/a^3$$

$$1.163 < a \quad \varepsilon_{2.4}(a) = 2.20 \times 10^{34}/a^3$$

Here, $a^2 = r^2 + z^2/(1-e^2)$, $e^2 = 0.4375$
where r and z represent the cylindrical coordinates with the origin at the galactic center in kpc unit and e is an eccentricity of the spheroid.

The surface brightness distribution derived from the model is compared with observation in Fig. 3 in which the data points by Becklin and Neugebauer[3] after correcting for $A_v$=27 mag are included.

## MASS TO LUMINOSITY RATIO

Since the 2.4 μm volume emissivity is closely related to the mass distribution, our model must provide some constraints on dynamical model based on HI observation. Assuming that the mass to luminosity ratio, M/L and color are constant, we compare the rotation velocity curve obtained by Sinha[14] with the circular velocity expected from our model in Fig. 4. The circular velocity curve calculated for M/L=2.0 (solar unit) and the color of the central bulge of M31[15], well represents the observed velocity for the southern galactic plane, whereas agreement is poor for the northern galactic plane. Neglecting the fine structure, the average M/L for the central bulge is estimated to be $2.0\pm0.4$. This value is somewhat smaller than the value used by Sanders and Lowinger[16] and significantly smaller than 10 observed for M31[17]. The latter implies that our central bulge emits radiation more efficiently.

Although our model qualitatively explains the observed velocity distribution, the latter deviates from that calculated with our model in fine structure such as the south-north asymmetry. This is probably caused by the non-uniform density distribution and/or non-circular motion of gas in the bulge.

## CONCLUSION

The model of the distribution of 2.4 μm volume emissivity in the central bulge was obtained based on the balloon observation of the surface brightness at 2.4 and 3.4 μm. The interstellar extinction of $A_V=15\pm4$ mag is obtained for the region within 1° from the center. A low M/L, $2.0\pm0.4$, was obtained by comparing the observed rotation velosity with the circular velocity expected from the model. Near infrared observation provides constraints on the model of the bulge and the non-uniform density and velocity distribution of gas.

## REFERENCES

1. R.J. Cohen and R.D. Davies, Mon. Not. R. Astr. Soc., **175**, 1 (1976)
2. H.S. Liszt and W.B. Burton, Astrophys. J., **236**, 779 (1980)
3. E.E. Becklin and G. Neugebauer, Astrophys. J., **151**, 145 (1968)
4. K. Ito, T. Matsumoto and K. Uyama, Nature, **265**, 517 (1977)
5. H. Okuda, T. Maihara, N. Oda and T. Sugiyama, Nature, **265**, 515 (1977)
6. W. Hofmann, D. Lemke and C. Thum, Astron. Astrophys. **57**, 111 (1977)
7. S. Hayakawa, T. Matsumoto, H. Murakami, K. Uyama, T. Yamagami and J.A. Thomas, New Zealand J. Sci., **22**, 353 (1979)
8. T. Ichikawa, K. Hamajima, K. Ishida, B. Hidayat and M. Raharto, submitted to Publ. Astron. Soc. Japan
9. R.L. Dickman, Astrophys. J. Suppl., **37**, 407 (1978)
10. S. Hayakawa, T. Matsumoto, H. Murakami, K. Uyama, J.A. Thomas and T. Yamagami, Astron. Astrophys., **100**, 116 (1981)
11. E.E. Becklin, K. Matthews, G. Neugebauer and S.P. Willner, Astrophys. J. **220**, 831 (1978)
12. J.H. Oort and L. Plaut, Astron. Astrophys., **41**, 71 (1975)
13. W.E. Harris, Astron. J., **81**, 1095 (1976)
14. R.P. Sinha, Astron. Astrophys., **69**, 227 (1978)
15. A.R. Sandage, E.E. Becklin and G. Neugebauer, Astrophys. J., **157**, 55 (1969)
16. R.H. Sanders and T. Lowinger, Astron. J., **77**, 292 (1972)
17. T.B. Williams, Astrophys. J., **214**, 685 (1977)

Chapter III     Infrared Spectroscopy

# INFRARED OBSERVATIONS OF THE IONIZED GAS IN THE GALACTIC CENTER

J. H. Lacy
California Institute of Technology, Pasadena CA 91125

## ABSTRACT

Observations of ionic emission lines from the Galactic center are discussed. These observations provide information regarding the abundances of heavy elements, the spectrum of ionizing radiation, the distribution of ionized gas, and the gravitational potential in which the gas moves. There is evidence from the gas cloud velocities that a massive central black hole is present, but other considerations do not support this conclusion. The most plausible scenario involves a dense central star cluster in which stellar collisions disrupt stars, and recently formed OB stars provide the radiation which ionizes the gas.

## I. INTRODUCTION

The first observations of an ionic emission line from the Galactic center were made by Aitken, Jones, and Penman,[1] who detected the 12.8 $\mu$m line of [Ne II]. These observations demonstrated the thermal nature of the radio continuum emission from Sgr A West and indicated that neon is predominantly singly ionized in the H II region unless it is substantially overabundant there. [Ne II] has subsequently been observed by Aitken et al,[2] Wollman et al,[3,4] and Lacy et al.[5,6] Wollman et al resolved the line spectrally, finding a linewidth of ~200 km s$^{-1}$, which indicates a virial mass of ~4×10$^6$ M$_\odot$ within a 1 pc radius. Lacy et al resolved the line spatially as well, and found that the ionized gas is concentrated in at least 14 clouds, each with a different velocity.

Emission lines of several other ions have also been observed and provide important information on the atomic abundances and the excitation of the ionized gas. Brackett and Pfundt lines of H I have been observed by Soifer et al,[7] Neugebauer et al,[8] Bally et al,[9] Lester et al,[10] and Nadeau et al.[11] [Ar II] (7.0 $\mu$m) has been observed by Willner et al[12] and Lester et al.[10] Lacy et al[5,6] searched for [Ar III] (9.0 $\mu$m), [S IV] (10.5 $\mu$m), and [Ar V] (13.1 $\mu$m), and McCarthy et al[13] searched for [S III] (18.7 $\mu$m). Of these four lines, only [Ar III] was detected, and it only marginally. [O III] (51.8 $\mu$m and 88.4 $\mu$m) have been observed by Dain et al[14], and Watson et al,[15] and Lester et al[16] have observed [O I] (63.2 $\mu$m).

In § II we discuss conclusions from the infrared ionic line observations regarding the atomic abundances and the excitation, distribution, and dynamics of the gas. We show that in several respects the Galactic center differs significantly from other H II regions. In § III we discuss models of the Galactic center. We first discuss the possibility that the presence of a massive central black hole could explain the unusual aspects of the observations. We conclude that although its presence cannot be ruled out, only the weak dynamical evidence favors the black hole hypothesis. An alternative model is discussed in § III C.

0094-243X/82/830053-07$3.00 Copyright 1982 American Institute of Physics

## II. OBSERVATIONS

### A. ATOMIC ABUNDANCES

Optical observations indicate the presence of gradients in z, the abundance of heavy elements, in several external galaxies and between ~8 and 15 kpc in the Galaxy, with z increasing toward the center. Radio observations of electron temperatures in Galactic H II regions provide indirect evidence of a continuation of the abundance gradient in to at least 5 kpc. A measurement of z in the Galactic center would provide important information about the amount of stellar processing which has ocurred there.

Neon and argon are the only heavy elements for which the dominant ionization states have been observed in the Galactic center. The apparent ratio of $Ne^+/H^+$ is approximately equal to the solar Ne/H ratio.[6] There are several problems with this measurement, however. The extinction correction is uncertain by ± 0.5 mag, neutral neon may be present, and collisional deexcitation of the fine-structure transition may be important. All three problems are less severe for argon. Both $Ar^+$ and $Ar^{+2}$, which cover the range 16 - 41 eV in excitation, have been observed so correction for unobserved ionization states is probably negligible. In addition, H I Pfα (7.5 μm) was observed simultaneously with [Ar II] (7.0 μm), allowing a reddening free measurement of $Ar^+/H^+$.[10] The observed $(Ar^+ + Ar^{+2})/H^+$ ratio is about twice the solar value, with the primary uncertainties being due uncertainties in the electron temperature and the fine-structure collision strength.

### B. EXCITATION OF THE GAS

The ratios of the fine-structure line fluxes indicate that $Ar^{+2}/Ar^+ \sim 0.1$ and $S^{+3}/S \lesssim 0.02$, assuming in the latter case that S/H equals the solar value. In contrast, most H II regions have $Ar^{+2}/Ar^+ \gtrsim 1$, and $S^{+3}/S$ often approaches unity. The most plausible explanation which has been suggested for the excitation of the gas is photoionization by ultraviolet radiation with a softer spectrum than is typical of that from stars which ionize H II regions in the solar neighborhood. If the ionizing radiation is emitted by a single central source, the spectral distribution would have to be like that of a star with $T_{eff} \lesssim 31{,}000K$ (spectral type O9.5). If ionizing sources are distributed throughout the central parsec, they could have effective temperatures up to ~35,000K (O8.5). A total ionizing luminosity of $N_{Lc} \sim 4 \times 10^{50}$ s$^{-1}$ is required to ionize the observed gas. If the ionizing sources have spectra like those of O8.5 V stars, their bolometric luminosities would be ~$2 \times 10^7$ L$_\odot$.

### C. DISTRIBUTION OF THE GAS

Until recently, lack of sensitivity in the infrared and a suitable radio interferometer made it impossible to measure the distribution of the ionized gas directly. However, 10 μm continuum emission from warm dust has been found to be a good tracer of ionized gas since intense radiation fields are necessary to heat dust sufficiently to allow it to radiate in the mid-infrared. Maps of the 10 μm continuum radiation show a complicated distribution of the warm dust, suggesting that the ionized gas is also very clumpy. This suggestion has been confirmed by observations of [Ne II][6] and radio free-free emission.[17] A map of the 10 μm continuum radiation from the central arc-minute,[18] with representative [Ne II] spectra superposed, is shown in Figure 1. The [Ne II] emission has been shown to follow the 10 μm continuum quite closely, with each peak in the continuum map seen at a different velocity in the line. In all, 14 distinguishable clouds of ionized gas have been seen. These clouds typically have densities of $n_e \sim 10^5$ cm$^{-3}$, diameters of $D \sim 0.2$ pc, ionized masses of $M \sim 1$ $M_\odot$, internal velocity dispersions of $\Delta v \sim 100$ km s$^{-1}$, and each absorbs a flux of ionizing radiation of $N_{Lc} \sim 10^{49}$ s$^{-1}$. The clouds are probably immersed in a much lower density, possibly very hot, medium. No mechanism has been suggested to contain the clouds against their internal motions, implying that they can live for only $\sim 10^4$ yr, and new clouds must be created at a rate of $\sim 10^{-3}$ yr$^{-1}$.

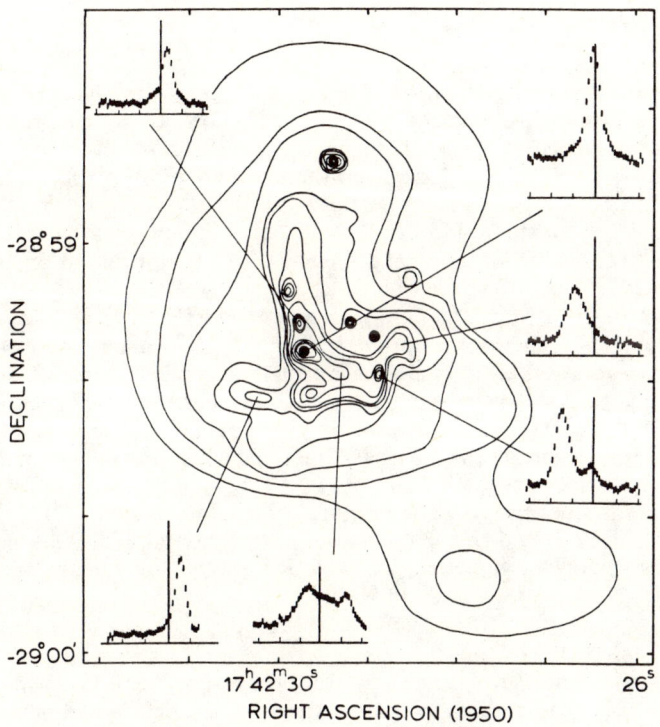

Fig. 1. 10 μm continuum map of the Galactic center[18] with representative [Ne II] spectra.[6] The central vertical axes of the spectra are at $v_{LSR} = 0$, and the horizontal axes are marked in 200 km s$^{-1}$ intervals.

## D. DYNAMICS

If gravity is the dominant force acting on the gas clouds, the clouds can be used as test masses to measure the mass distribution in the Galactic center. The best fit mass distribution includes a central mass of $\sim 3\times 10^6$ $M_\odot$ in addition to a distributed (stellar) mass of $\sim 3\times 10^6$ $M_\odot$ in the central pc. However, because of the small number of test masses, the statistical significance of this mass determination is not large; a distributed mass of $\sim 10^7$ $M_\odot$ with no central massive object is allowed at the $1.4\sigma$ level. In addition, the possibility of acceleration of the clouds by non-gravitational forces, which would invalidate their use as test masses, should be kept in mind.

## III. DISCUSSION

### A. PROBLEMS

The observations discussed above demonstrate that the Galactic center differs from other regions of ionized gas in the Galaxy in several important respects. The explanation of these differences is the primary problem in constructing a model of the region. The observational facts and the problems they present are:

1. The ionized gas is concentrated in at least 14 small clouds which can survive for only $\sim 10^4$ yr. A mechanism to generate clouds with the observed properties (see §II C) at a rate of $\sim 2\times 10^{-3}$ yr$^{-1}$ is required.
2. The gas is ionized by a radiation field with a spectral distribution like that of stars of $T_{eff} < 35,000$ K and with a luminosity of $\sim 4\times 10^{50}$ ionizing photons per second. For this radiation to be supplied by main-sequence stars, $\sim 200$ O9 V stars would be required, with essentially no stars of earlier spectral types.
3. The clouds are dissipated either by colliding with each other or by expansion at the rate determined from the observed internal velocity dispersions. The disruption of the clouds generates $\sim 10^{-3}$ $M_\odot$ yr$^{-1}$ which must be removed from the interstellar medium of the Galactic center.
4. The cloud velocities, if due to orbital motions in a gravitational field, suggest the presence of a $\sim 3\times 10^6$ $M_\odot$ object at the Galactic center.

### B. BLACK HOLE MODEL

Because the dynamical evidence suggests the presence of a massive object at the Galactic center it is natural to ask whether a black hole would have other observational consequences, especially regarding the other problems presented by the observations.[19] Three such consequences seem most relevant. First, the radiation from an accretion disk surrounding a black hole could account for the unusual excitation of the gas or could be observable directly. Second, the gas clouds could result from stars which are tidally disrupted during near encounters with the black hole. And third, the disrupted clouds could either accrete into the black hole or be swept away by radiation pressure from the accreting gas.

Many models have been proposed for the radiation emitted by gas accreting into a black hole. We will consider just one of these, the "α model" for disk accretion.[20] Disk accretion provides a spectrum closer to that required than does spherical accretion and seems more likely to apply to the Galactic center since the observed gas appears to have significant angular momentum. In the simplest approximation, each point on the disk radiates as a blackbody with a temperature sufficient to radiate the energy deposited at that radius as gas spirals in toward the center. The resulting spectrum resembles a blackbody spectrum, but with a long wavelength tail falling as $\nu^{1/3}$ rather than $\nu^2$. The wavelength of the peak flux is determined by the maximum effective temperature of the disk which decreases as the mass of the black hole increases. For a black hole of $3\times10^6\,M_\odot$ accreting at a rate of $2\times10^{-5}\,M_\odot\,\text{yr}^{-1}$, the short wavelength edge of the accretion disk spectrum matches that of a blackbody of ~32,000 K. Because the accretion disk is not an isothermal gray slab at each radius, its spectrum does not match that predicted by this simple calculation. However, since the surface gravity, and presumably the chemical composition of the disk is quite similar to that of stars, the disk spectrum will resemble that of a distribution of stars of various effective temperatures. The far ultraviolet radiation will then have a spectral distribution like that of an O star of $T_{eff} \sim$ 32,000K, and the ionizing spectrum of the disk should be essentially that required to ionize the gas clouds. A problem arises, however, because the $\nu^{1/3}$ tail of the accretion disk spectrum would be ~ 100 times as bright as any observed source at 2.2 $\mu$m. The 2.2 $\mu$m flux of the disk would be decreased if the disk could be cut off at a radius inside of that at which most of the 2.2 $\mu$m flux is emitted. However, even a spectrum like that of a 32,000 K star would be too luminous at 2.2 $\mu$m if all of the ionizing radiation is emitted by a single source. This argument does not rule out the presence of a massive black hole, but it would have to accrete gas at a rate too low to account for the ionization of the gas clouds.

Two matters must be considered to determine whether tidal disruption of stars could account for the generation of the gas clouds: the rate of tidal disruptions and the characteristics of the resulting gas. In the first respect this mechanism appears to be promising. For the best estimates of the parameters which describe the black hole and the stellar density, a disruption rate of ~$10^{-3}\,\text{yr}^{-1}$ is calculated.[19,21] This rate agrees quite well with that required for cloud production. The appearance of the disrupted stars has not been calculated in any great detail, but the velocity dispersion of the disrupted stars can be estimated from the difference in potential energy across a star at the point at which the black hole tidal force overcomes the stellar surface gravity. Such a calculation indicates that about half of the star is ejected with velocities up to ~ 5000 km s$^{-1}$, with the remaining gas forming a tightly bound disk around the black hole. Neither part of the disrupted star resembles the observed clouds. In addition, the average luminosity of the accreting stars is $\gtrsim 10^8\,L_\odot$, comparable to the total luminosity which is observed from the central 100 pc. In order to avoid violating observational constraints on the luminosity of an object at the Galactic center, it must be assumed that the gas accretes very rapidly into the black hole and that we are observing a lull between bursts of luminosity.

A black hole would provide a means of removing the gas from the disrupted clouds if it could accrete into the hole. However, accretion of ~$10^{-3}\,M_\odot\,\text{yr}^{-1}$ by disk accretion would emit much more radiation than is observed. The observed 2.2 $\mu$m luminosity limits the present disk accretion rate to $\lesssim 10^{-6}\,M_\odot\,\text{yr}^{-1}$. Spherical accretion is much less efficient at generating luminosity so that an accretion rate of $10^{-3}\,M_\odot\,\text{yr}^{-1}$ would be permitted, but for the reasons given above disk accretion seems much more likely. If any significant quantity of gas accretes into a black hole, it probably does so sporadically, with the present accretion rate well below the average. Radiation pressure may be important if the gas accretes sporadically. A rapid accretion of $\sim 1 M_\odot$ would provide sufficient radiation pressure to sweep any gas out of the central parsec.

## C. STAR CLUSTER MODEL

We have argued above that a massive black hole cannot provide the ionizing radiation which is absorbed in the central parsec, and that tidal disruptions of stars which pass near a black hole cannot produce the gas clouds. We now discuss processes which could occur in the absence of a black hole which could produce and ionize the gas clouds, and discuss whether all observations described above can be explained in the context of a star cluster with no central massive black hole.

The most promising cloud formation mechanism involves disruption of stars by binary stellar collisions.[19] If the stellar distribution is that of an isothermal cluster with $\sigma = 126$ km s$^{-1}$ (the velocity dispersion of the gas clouds), and a core radius $\lesssim 0.03$ pc, the collision rate between main-sequence stars would be $\sim 10^{-3}$ yr$^{-1}$. These collisions would not, however, produce gas clouds like those seen. Too little gas would be liberated, and that which is would have too high a velocity dispersion. Collisions between red giants could produce clouds like those seen, but the collision rate is much too low. On the other hand, collisions between main-sequence and red giant stars may be sufficiently frequent and may produce enough gas with the observed dispersion; an accurate calculation would be very useful. No model seems entirely satisfactory, and other processes to generate, or perhaps regenerate, the clouds should be studied.

There are several ways to alter the spectrum of an OB asociation to match that required to ionize the Galactic center. If stars could form with an initial mass function which strongly favors stars of 20 - 30 $M_\odot$, or if the relation between stellar mass and effective temperature could be different from that seen in the solar neighborhood, a suitably soft spectrum could result. Either of these circumstances could result from the metal rich interstellar medium which is probably present at the Galactic center, and there is evidence from observations of other galaxies that one of these processes occurs toward galactic nuclei. The low excitation of the Galactic center could also be explained if it is ionized by an OB association which formed $\sim 4 \times 10^6$ years ago. Because the most massive stars cool rather quickly as they evolve, such a cluster would now be characterized by a spectrum $\sim$ O9.

In the absence of a black hole which could accrete or blow away the disrupted gas clouds, the most likely sink for the gas is formation of stars. These stars may or may not provide the ionizing radiation, depending on the masses of stars which are formed. Presumably the gas would have to collect at the center of the central star cluster before it could collapse to form stars. As no very massive collection of gas is seen in the ionic emission lines, if present it must be sufficiently dense to shield itself from the ionizing radiation and so be mostly neutral.

## IV. SUMMARY

Observations of infrared emission lines provide some of the most important information regarding the content of the central parsec of the Galaxy. Perhaps the most interesting conclusion from these observations is the dynamical evidence of the presence of a massive central black hole. In this review, other aspects of the observations have been discusssed, to try to confirm or deny this suggestion. No further evidence of a massive black hole is found, and the observations seem to be explained best by processes which are expected in a very dense star cluster in which there has been recent OB star formation.

## REFERENCES

1. D. K. Aitken, B. Jones, and J. M. Penman, M.N.R.A.S, **169**, 35P (1974).
2. D. K. Aitken, J. Griffiths, B. Jones, and J. M. Penman, M.N.R.A.S, **174**, 41P (1976).
3. E. R. Wollman, T. R. Geballe, J. H. Lacy, C. H. Townes, and D. M. Rank, Ap. J. (Letters), **205**, L5 (1976).
4. ---------, Ap. J. (Letters), **218**, L103 (1977).
5. J. H. Lacy, F. Baas, C. H. Townes, and T. R. Geballe, Ap. J. (Letters), **227**, L17 (1979).
6. J. Lacy, C. H. Townes, T. R. Geballe, and D. J. Hollenbach, Ap. J., **241**, 132 (1980).
7. B. T. Soifer, R. W. Russell, and K. M. Merrill, Ap. J. (Letters), **207**, L83 (1976).
8. G. Neugebauer, E. E. Becklin, K. Matthews, and C. G. Wynn-Williams, Ap. J., **220**, 149 (1978).
9. J. Bally, R. R. Joyce, and N. Z. Scoville, Ap. J., **229**, 917 (1979).
10. D. F. Lester, J. D. Bregman, F. C. Witteborn, D. M. Rank, and H. L. Dinerstein, Ap. J., **248**, 524 (1981).
11. D. Nadeau, G. Neugebauer, K. Matthews, and T. R. Geballe, A. J., **86**, 561 (1981).
12. S. P. Willner, R. W. Russell, R. C. Puetter, B. T. Soifer, and P. M. Harvey, Ap. J. (Letters), **229**, L65 (1979).
13. J. F. McCarthy, W. J. Forrest, D. A. Briotta, and J. R. Houck, Ap. J., **242**, 965 (1980).
14. F. W. Dain, G. E. Gull, E. Melnick, and M. Harwit, Ap. J. (Letters), **221**, L12 (1978).
15. D. M. Watson, J. W. V. Storey, C. H. Townes, and E. E. Haller, Ap. J. (Letters), **241**, L43 (1980).
16. D. F. Lester, M. W. Werner, J. W. V. Storey, D. M. Watson, and C. H. Townes, Ap. J. (Letters), **248**, L109, (1981).
17. R. L. Brown, K. J. Johston, and K. Y. Lo, Ap. J., **250**, 155 (1981).
18. G. H. Rieke, C. M. Telesco, and D. A. Harper, Ap. J., **220**, 556 (1978).
19. J. H. Lacy, C. H. Townes, and D. J. Hollenbach, submitted to Ap. J. (1981).
20. I. D. Novikov and K. S. Thorne, in **Black Holes**, ed. C. DeWitt and B. DeWitt, p. 343 (1973).
21. A. B. Marchant and S. L. Shapiro, Ap. J., **239**, 685 (1980).

## Bα AND Ne II LINE SPECTROSCOPY IN THE VICINITY OF THE GALACTIC CENTER SOURCE IRS 16

T.R. Geballe and S.E. Persson
Mt. Wilson and Las Campanas Observatories
Carnegie Institution of Washington, Pasadena CA

J.H. Lacy and G. Neugebauer
Palomar Observatory
California Institute of Technology, Pasadena CA

S.C. Beck
University of California, Berkeley CA

### ABSTRACT

Bα (4.05 $\mu$m) and [Ne II] (12.81 $\mu$m) line spectra at spatial resolutions of $\sim 3"$ and spectral resolutions of 80 km s$^{-1}$ have been obtained on a grid of positions surrounding IRS 16, which may be at the Galactic center. The Bα and [Ne II] line profiles agree within the uncertainties and neither set of spectra shows evidence of ionized gas associated with IRS 16 over a velocity range of $-500$ to $+450$ km s$^{-1}$. A spectrum of Bα from an 8" beam centered on IRS 16 and covering $-7500$ to $+6500$ km s$^{-1}$ with 500 km s$^{-1}$ resolution shows marginal evidence for a broad-line component. These data imply that no concentration of moderate velocity gas ($|v| < 300$ km s$^{-1}$) is directly associated with IRS 16 and that the ionized gas near IRS 16 is for the most part neither very dense nor of very high velocity.

### INTRODUCTION

The nature of the peculiar object IRS 16, which lies at or very near the center of the Galaxy and which may be associated with an ultracompact radio source, is uncertain. IRS 16 has been spatially resolved in the near infrared continuum and lies near the center of the extended near infrared emission, suggesting that it is the central stellar concentration of the Galaxy.[1] However, absorption due to the 2.3 $\mu$m CO bands is not seen at the level expected in a star cluster.[2] Bα, Bγ, and [Ne II] emission lines are seen toward IRS 16,[3-7] but no definite relationship of that source with any particular velocity component of the ionized gas has been established. If IRS 16 is the core of a central star cluster, gas collecting there would be expected to emit symmetric lines with widths less than or comparable to that of the surrounding region, $\sim 300$ km s$^{-1}$. Alternatively, if a massive object is present at the center, broader lines would be expected. In addition, if the ionized gas near IRS 16 were very dense, [Ne II] and other fine structure lines might be suppressed.

0094-243X/830060-07$3.00 Copyright 1982 American Institute of Physics

Previous studies have not had sufficient angular resolution to separate an ionized gas component associated with IRS 16 from nearby clouds and have observed too small a range of velocity to detect a line as broad as those seen, for example, from Seyfert galaxies. In an attempt to improve on them, we have obtained closely spaced grids of B$\alpha$ and [Ne II] spectra with angular resolutions of $\sim 3"$ and spectral resolutions of 80 km s$^{-1}$ to search for moderate velocity dispersion gas associated with IRS 16 and we have obtained a lower resolution spectrum covering a wide velocity range to search for a broad-line component of B$\alpha$.

## OBSERVATIONS

The high resolution B$\alpha$ spectra, shown in Figure 1, were obtained with a Fabry-Perot and cooled grating spectrometer[8] at the 5m Hale telescope at Palomar Observatory during 1981 July. A 2".5 focal plane aperture and a spectral resolution of 80 km s$^{-1}$ were used. IRS 16 was centered in the beam by first offsetting from IRS 7 and then maximizing the 2$\mu$m signal. A B$\alpha$ spectrum was obtained at this position and at a hexagonal pattern of positions, each one 2" from IRS 16.

The [Ne II] spectra, shown in Figure 2, were obtained during 1979 May at the 2.5m du Pont telescope of Las Campanas Observatory, using a cooled Fabry-Perot and grating spectrometer.[9] A spectral resolution corresponding to 80 km s$^{-1}$ and a 3" aperture were used. Spectra were taken on a rectangular grid, with positions spaced by 0$^s$.15 in RA and 2" in Dec. The positions were measured by offsetting from a visual field star at $17^h 42^m 30^s.0$, -28°59'01" (1950). IRS 16 lies approximately half way between the positions of spectra 6 and 7.

The low resolution B$\alpha$ spectrum, in Figure 3, was measured in 1981 April at the du Pont telescope with a stepping grating spectrometer.[8] The spectrum was taken through an 8" aperture centered on the 2$\mu$m continuum peak of IRS 16, and covered -7500 to +6500 km s$^{-1}$ with 500 km s$^{-1}$ resolution.

Cumulative positional uncertainties due to offsetting and guiding were $\sim$1" for all observations and the seeing disks were in all cases smaller than the apertures used. Absolute flux calibrations should be accurate to $\pm$20 percent; and within each grid relative fluxes are accurate to $\pm$5 percent.

## B$\alpha$ AND [Ne II] HIGH RESOLUTION SPECTRA

The B$\alpha$ and [Ne II] spectra in Figures 1 and 2 show emission over at least -300 to +250 km s$^{-1}$ as well as rapid variations in the line intensities and profiles. The lines increase in width to the southwest and increase in strength to the south and east of IRS 16. These variations are apparently due to the presence of several clouds in the vicinity of IRS 16.[6] It appears, however, that no cloud is centered on IRS 16 itself; there is no obvious

component of the line toward IRS 16 in this velocity range which
cannot be attributed to nearby clouds. To quantify this
conclusion we have compared the average of the six B$\alpha$ spectra
surrounding IRS 16 with the central spectrum. The uppermost panel
in Figure 1 shows the difference between IRS 16 and the average of
its surroundings, plotted on the same scale as the original
spectra. It is clear that no velocity component peaks on IRS 16.
The total intensities (line plus continuum) integrated over the
velocity range -500 to +450 km s$^{-1}$, of the spectrum of IRS 16 and
that of the surrounding region agree to better than 4 percent.
Since the ionized gas clouds in the Galactic center are extended,
with sizes of typically 4", it seems probable that the emission
seen toward IRS 16 is due to nearby clouds which overlap the line
of sight to IRS 16.

A similar comparison of the [Ne II] line profile with that of
the surrounding region was also made. Again no component of the
line clearly associated with IRS 16 was seen. There is some
evidence in the [Ne II] spectra, which cover a wider area than
those of B$\alpha$, that features shift toward the blue in going to the
west or southwest. It is difficult to say whether this is due to
rotation of extended clouds of gas or simply overlapping red- and
blue-shifted clouds. In any case it is not possible to choose a
center of rotation unambiguously. As in the B$\alpha$ spectra, the
emission toward IRS 16 appears to be a continuation of the pattern
seen near it, and does not show a separate gas concentration.

Uncertainty in the relative positioning of the B$\alpha$ and
[Ne II] spectra make a detailed comparison of the two lines
difficult. In general, however, they agree quite well. For
example, note the double-peaked lines to the southwest of IRS 16
and the overall increase in intensity toward the southeast. The
intensity gradient also agrees well with the structure of the
region at 10$\mu$m. This indicates that in the velocity range
-500 to + 450 km s$^{-1}$ there is no dense ($n_e > 3 \times 10^5$ cm$^{-3}$) ionized
gas which would produce the B$\alpha$ emission but not the forbidden
[Ne II] line.

## B$\alpha$ LOW RESOLUTION SPECTRUM

The 500 km s$^{-1}$ resolution B$\alpha$ spectrum (Figure 3) was fitted
with a Gaussian line profile plus a linear continuum. The line
fit, which has a FWHM of 600 km s$^{-1}$, is consistent with the B$\alpha$
line width found from the higher resolution observations reported
above. There is evidence for a slight excess of emission centered
at $\sim$ 4000 km s$^{-1}$ on either side of the narrow line, but this may
be due to instrumental effects. However, a spectrum of the bright
compact HII region G333.6-0.2 has a flat continuum when the same
data reduction technique is applied. It can be concluded
nonetheless that any broad-line component near IRS 16 has less
than a few percent of the flux density of the narrow component and
that the flux of a broad component, integrated over -7500 to +6500

km s$^{-1}$, is significantly less than that of the narrow line in an 8" beam.

## THE 4 μm CONTINUUM

Five of the Bα spectra in Figure 1 show a strong continuum at 4 μm. The flux density of the continuum agrees with that of Becklin et al[10], to within the uncertainties imposed by the differences in aperture size (3".8 versus 2".5), pointing, and photometric bandwidths used. We note that the equivalent width of

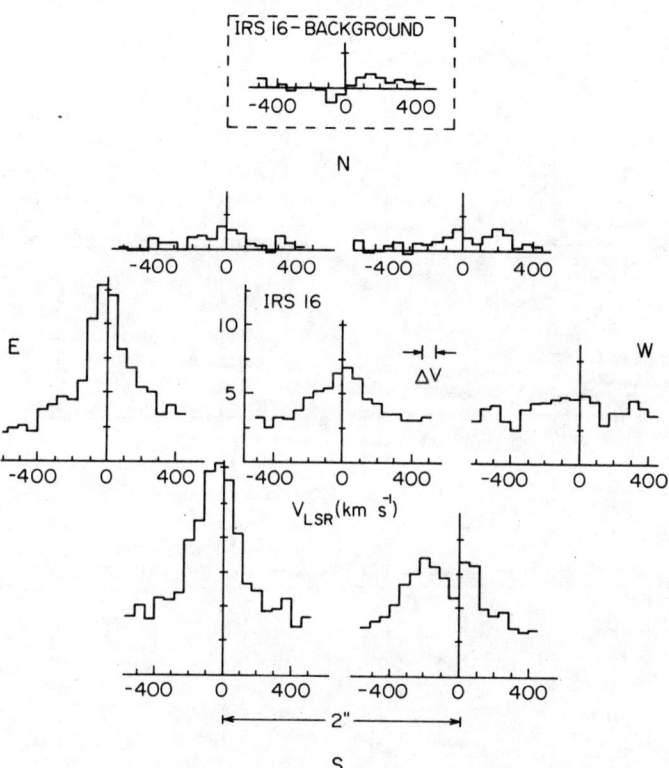

Fig. 1. Spectra of Bα (4.05 μm) near IRS 16. The data are binned in 57.5 km s$^{-1}$ intervals, and the velocity resolution is 80 km s$^{-1}$. The flux scales are labeled in units of $10^{-20}$ W/cm$^2$ cm$^{-1}$ in the 2".5 diameter beam. The central spectrum was taken at the position of IRS 16; the spectrum enclosed by dashed lines is the difference of the central spectrum and the average of the six surrounding spectra.

Bα changes by at least a factor of 10 in the east-west direction across IRS 16, while much less change is seen in the [Ne II] spectra of Figure 2. Such a change could be due to rapid changes in the dust temperature or ionization, but is most easily interpreted as arising from additional sources of non-dust (e.g. stellar) sources of continuum radiation. It seems likely then that the excess 4 μm continuum which is seen in the central, W, and SW grid spectra is due to IRS 16 itself. More detailed mapping measurements would be necessary to resolve IRS 16 out of the 4 μm background.

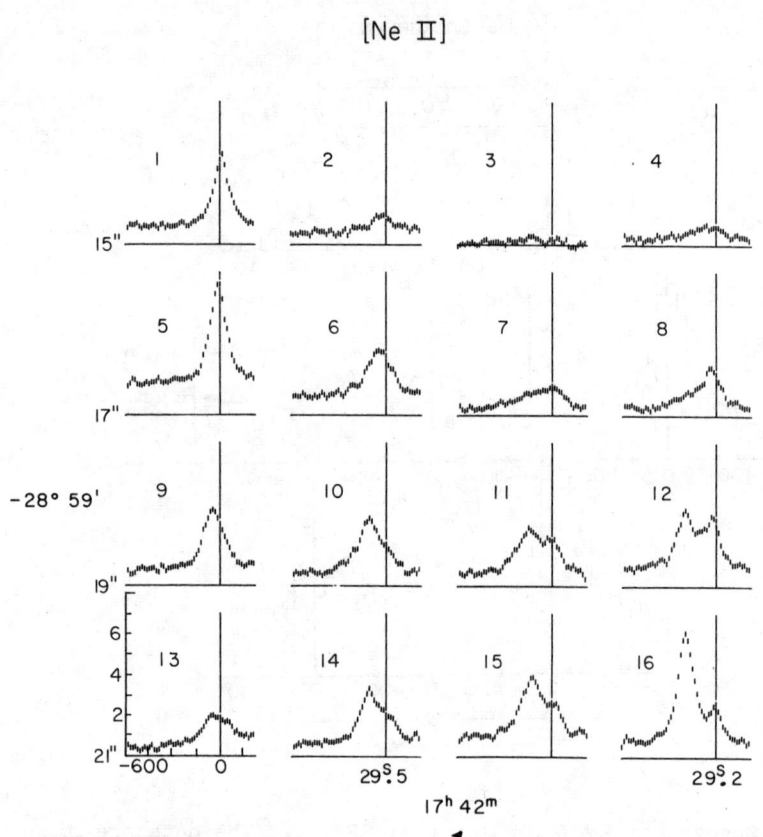

Fig. 2. Spectra of [Ne II] (12.81 μm) near IRS 16. The data are binned in 20 km s$^{-1}$ intervals, and the velocity resolution is 80 km s$^{-1}$. The intensity scale is labeled in units of $10^{-8}$ W/cm$^2$ cm$^{-1}$ sr. IRS 16 lies approximately half way between spectra 6 and 7.

## SUMMARY

1. No feature which is centered spatially on IRS 16 is seen in either B$\alpha$ or [Ne II] between -500 and +450 km s$^{-1}$.
2. Over the region mapped, no feature is seen in B$\alpha$ which is not also seen in the [Ne II] forbidden line.
3. A broad-line component of B$\alpha$ in the range -7500 to +6500 km s$^{-1}$ is at best only a minor part of the line emission seen toward IRS 16.

These conclusions indicate that whatever the nature of IRS 16 is, relatively little or no ionized gas, whether of high or low density and high or low velocity dispersion, is associated with its vicinity.

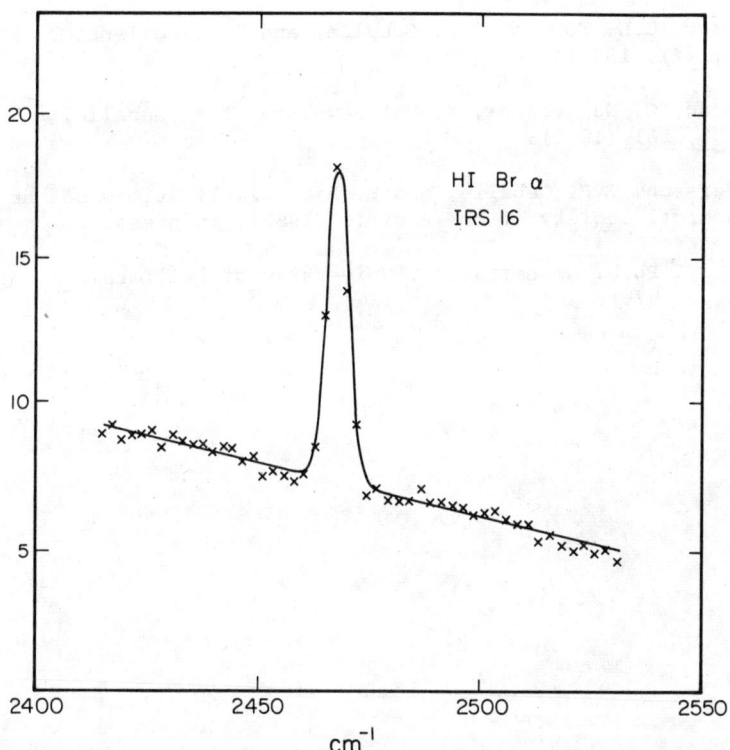

Fig. 3. Spectrum of B$\alpha$ from an 8" beam centered on IRS 16. Data points are separated by 286 km s$^{-1}$, and the velocity resolution is 500 km s$^{-1}$. The flux scale is labeled in units of $10^{-20}$ W/cm$^2$ cm$^{-1}$. Flux calibration and removal of telluric and instrumental features were obtained from a comparison star. The smooth curve is the best fit Gaussian line plus linear continuum.

## REFERENCES

1. E.E. Becklin, K. Matthews, G. Neugebauer, and S.P. Willner, Ap. J., $\underline{219}$, 121 (1978).

2. K.Y. Lo, J.H. Lacy, T.R. Geballe, and S.E. Persson, in preparation.

3. G. Neugebauer, E.E. Becklin, K. Matthews, and C.G. Wynn-Williams, Ap. J., $\underline{220}$, 149 (1978).

4. J. Bally, R.R. Joyce, and N.Z. Scoville, Ap. J., $\underline{229}$, 917 (1979).

5. J.H. Lacy, F. Baas, C.H. Townes, and T.R. Geballe, Ap. J. (Letters), $\underline{227}$, L17 (1979).

6. J.H. Lacy, C.H. Townes, T.R. Geballe, and D.J. Hollenbach, Ap. J., $\underline{241}$, 132 (1980).

7. D. Nadeau, G. Neugebauer, K. Matthews, and T.R. Geballe, A.J., $\underline{86}$, 561 (1981).

8. S.E. Persson, T.R. Geballe, and F. Baas, Publications of the Astronomical Society of the Pacific (1982), in press.

9. J.H. Lacy, Ph.D. dissertation, University of California, Berkeley (1979).

# SPATIAL AND SPECTRAL STUDIES OF THE GALACTIC CENTER NEAR 10μm

David K. Aitken*
Anglo Australian Observatory, Epping, NSW 2121

Michelle C. Allen
Sydney University, Sydney, NSW 2001

Patrick F. Roche
University College London, Gower St., London WC1E 6BT

## ABSTRACT

We present maps of part of SgrA (West), using a 3.5" diameter beam, in the continuum radiation at 10.5μm, at 12.5μm and in the NeII fine structure line at 12.81μm. There is a strong correlation between the infrared continuum sources and the peaks of the line emission, although the line emission has a more marked diffuse component. Many of the infrared sources are also positions of high color temperature, and it is argued that this implies that internal sources of luminosity power the discrete sources. There is no evidence for temperature structure at the position of the non-thermal radio source.

## INTRODUCTION

Whether the 10μm sources at the Galactic center are density enhancements excited by a single external source or contain local sources of luminosity, has been the subject of speculation. In the first case the distribution of ionization should be strongly correlated with the continuum emission because of nebular heating, and in second temperature gradients would be expected near the 10μm sources. Therefore a comparison of the distribution of nebular emission with that of the continuum and with color temperature should help to distinguish between these possibilities. So far a comparison of these quantities at adequate spatial and spectral resolutions has been lacking.

We present maps of SgrA (West) at 10.5 and 12.5μm and in the NeII line at 12.81μm obtained simultaneously with the same beam diameter of 3.5".

## OBSERVATIONS AND RESULTS

The observations were obtained using the UCL helium cooled array spectrometer at the f/36 chopping secondary focus of the 3.9m AAT in July 1981, with a beam size of 3.5" and beam throw 45" E-W.

---

* Present address: University College London, Gower St., WC1E 6BT

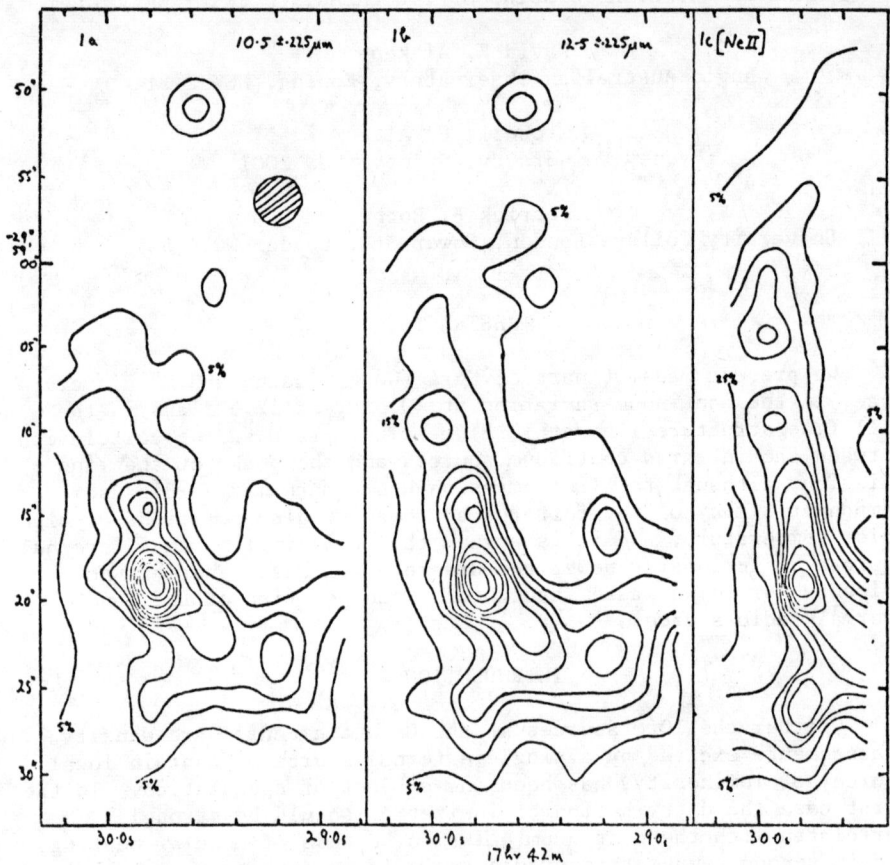

Figure 1. Intensity distributions at (a) 10.5μm, (b) 12.5μm, and (c) in the NeII line at 12.81μm, all with the same 3.5" diameter beam. All maps have been normalised to the same peak height and contour lines are at 10% intervals starting at 5%. X indicates the position of the non-thermal radio source.

Figures 1a and 1b show partial maps of SgrA in the continuum emission centered at 10.5μm and 12.5μm, each in a bandpass of 0.45μm. Figure 1c is a map in the NeII line intensity derived from the difference in response at 12.81μm in a bandpass of 0.025μm and the averaged response of channels covering the intervals 12.75-12.795 and 12.825-12.87μm. The continuum contamination of the NeII map is less than 10%. This spectral information was obtained through the same field stop at the same time and sampled every 0.5", along a set of N-S lines spaced 2.0" apart. Unfortunately the NeII map is truncated towards the west and omits the region of interest near the non-thermal radio source. For purposes of comparison all the maps are normalised to the same peak height. The peak values of the flux densities are $5.4 \times 10^{-7}$ and $13.5 \times 10^{-7}$ W cm$^{-2}$ μm$^{-1}$ sr$^{-1}$ and $3.4 \times 10^{-8}$ W cm$^{-2}$ sr$^{-1}$ for (a), (b) and (c) respectively.

It is clear that a high degree of correlation exists between the continuum and the NeII line intensities; in particular the NeII peak coincides closely with that of IRS1. Nevertheless, the NeII structure has a larger diffuse and ridgelike component than that of the continuum so that the correlation is not exact. Likewise the correlation between the two continuum maps is incomplete since the structure at 12.5μm is significantly more diffuse than at 10.5μm. All the maps show the most dominant structure of the broad band map of Becklin et al (1978).

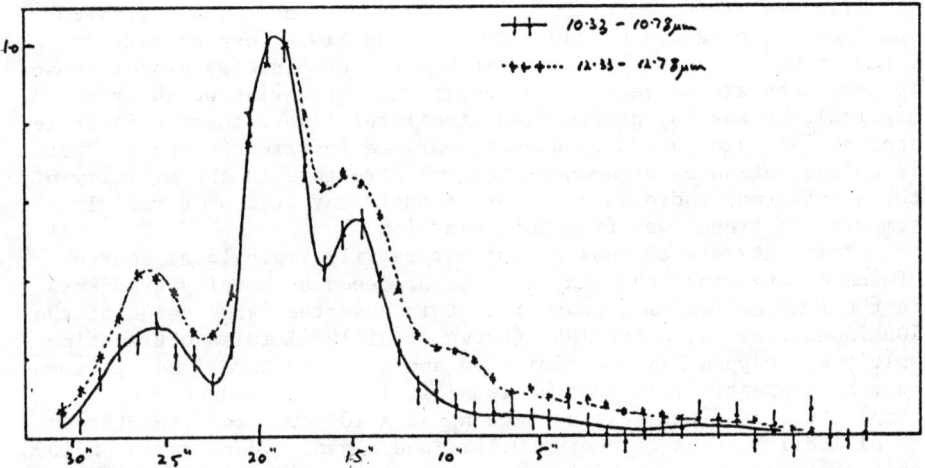

Figure 2. N-S scans through IRS1 at 10.5μm and 12.5μm. The peak heights have been normalised to unity. Errors are shown when they exceed 0.05 and are one standard deviation of the mean.

## DISCUSSION

We interpret the high degree of correlation between line emission and the dust continuum as evidence that the 10μm sources are density enhancements, as suggested by Lacy et al (1980). When we turn to the continuum maps at 10.5 and 12.5μm, we find that in general the shorter wavelength continuum is less extended and better resolves the discrete components. Figure 2 shows this more explicitly in the N-S scan at these two wavelengths through IRS1. We rule out the interpretation of this difference in terms of spatial variations of extinction since all the sources in the region, with the exception of IRS3 and IRS8, are known to have similar silicate absorption features (Willner 1978; Becklin et al 1978). Instead we interpret the difference, excepting these two sources, as due to temperature structure of the dust close to the discrete sources. An estimate of the grain temperature can be obtained by taking the emissivity of both the emitting and absorbing grains along the line of sight to be represented by the spectrum of the Trapezium in Orion (Forrest et al 1975). The emitting dust is taken to be optically thin and the optical depth in absorption at 9.7μm taken to be 3.5. In this way the observed colors are corrected for both extinction and emissivity and a grain temperature derived which varies from about 250 K at IRS1 to 150 K 3"S and roughly 10"N. Specifically grain temperature peaks are seen on IRS1, IRS2, IRS9 and IRS10 with values of 250, 200, 180 and 210 K respectively on a background of about 150 K. Although the absolute values of these temperatures are subject to systematic uncertainties which are difficult to assess, nevertheless their relative values do indicate substantial temperature gradients near the infrared sources. There is no indication of pronounced thermal structure in the vicinity of the non-thermal radio source, nor is there any sign of a radial temperature trend away from this position.

Temperature gradients do not necessarily imply local sources of luminosity since they may also be produced by density gradients in the ionized medium. However in this case the large value of the 10μm/5GHz flux ratio for IRS1 (Brown et al 1981) rules out heating solely by trapped Lyman α radiation and some additional mechanism must be present. This must be local if it is to produce a temperature gradient. If we take $n_e = 2 \times 10^4$ cm$^{-3}$ for IRS1 (Brown et al 1981), grains of radius 0.1μm, and a gas to dust ratio of 1000, then continuum heating will exceed Lyman α heating within 0.1 pc of a continuum source exceeding $10^6$ $L_\odot$. Up to 10 such sources would be within the total luminosity of $1-3 \times 10^7$ $L_\odot$ of the central region (Gatley, 1981) and this requirement on internal luminosity could be relaxed if the gas to dust ratio is larger and/or the grains are smaller.

We wish to acknowledge support for this work from the SERC and the assistance of the staff at the Anglo Australian Observatory.

### REFERENCES

Becklin, E.E., Mathews, K., Neugebauer, G., and Willner, S.P., Ap.J. 219, 121, 1978
Brown, R.L., Johnston, K.J., and Lo, K.Y. Ap.J., 250, 155, 1981
Forrest, W.J., Gillett, F.C., and Stein, W.A. Ap.J. 195, 423 1975
Gatley, I. (1981) IAU Symposium No.96, Ed. Wynn-Williams and Cruikshank, p.290

Lacy, J.H., Townes, C.H., Geballe, T.R., and Hollenbach, D.J.
    Ap.J. 241, 132, 1980
Willner, S.P., Ap.J. 219, 870, 1978

## OI AND OIII IN SGR A: NEUTRAL AND IONIZED GAS AT THE GALACTIC CENTER

R. Genzel, D. Watson, C. Townes
University of California, Berkeley, CA

D. Lester, H. Dinerstein, M. Werner
NASA Ames Research Center, Moffett Field, CA

J. Storey
AAT, Epping, NSW, Australia

### ABSTRACT

We have mapped the $^3P_1-^3P_2$ fine structure line emission at 63 microns from neutral oxygen in the vicinity of the galactic center. The emission is extended over more than 4' and is centered on Sgr A West. We conclude that the bulk of the OI emission arises in a predominantly neutral region outside of the ionized central 3 pc of our galaxy. Assuming that the oxygen is collisonally excited by neutral hydrogen impact, we estimate that the gas temperatures in this region are >130K, that is, significantly higher than the dust temperature of 70K. The OI line center velocities change systematically along the galactic plane in a manner consistent with galactic rotation. However, the unusual velocity distribution and linewidths suggest that the motions have a large noncircular component and that there are large scale inhomogeneities in the OI-emitting gas. We also have detected the 88 micron $^3P_1-^3P_0$ fine structure line of OIII in a 45" FWHM beam centered on Sgr A West. The ratio of this line intensity to that of the 52 micron $^3P_2-^3P_1$ line indicates that most of the ionized gas in this region has electron density $>10^4$ cm$^{-3}$.

### INTRODUCTION

Emission in the 63 micron OI fine structure line from the center of our galaxy was detected by Lester et. al.[1] We report here more extensive observations with better sensitivity and higher spectral resolution, to establish the angular distribution, kinematics and excitation of the OI emitting gas. In addition, we report the detection of the 88 micron $^3P_1-^3P_0$ line of OIII toward Sgr A and discuss its implication for the density distribution of ionized gas in that region.

### OBSERVATIONS

The data were taken from the NASA Kuiper Airborne Observatory in June of 1981. The LHe-cooled tandem scanned

Fabry-Perot spectrometer described by Storey, Watson and Townes[2] was used with a 45"-diameter beam. Resolving powers of 2000 and 1600 were used at 63 and 88 microns respectively. Absolute intensities were derived from the observed equivalent widths and the continuum fluxes from Gatley et. al.[3] in the same beam. The line was detected at 10 positions taken in 40" steps along and perpendicular to the galactic plane. The map center (0,0) was at (1950) $17^h 42^m 29^s$ -28° 59'. These spectra are shown in Figure 1 below.

## RESULTS

### 1. Line Fluxes

The peak intensity of the OI line in our 45" beam is $1.5\pm0.5 \times 10^{-16}$ W-cm$^{-2}$, in agreement with the value reported by Lester et. al.[1] The error in this measurement is primarily due to flux calibration uncertainties. The 88 micron OIII line was found to have an intensity of $7\pm2 \times 10^{-18}$ W-cm$^{-2}$ in the same beam. Our OIII measurement can be reconciled with the 4'x4.4' beam measurement by Dain et. al.[4] only if much of the emission detected in the larger beam comes from a region that is extended compared with our 45" beam.

### 2. Angular Distribution of OI Emission

The OI emission is extended over more than 4' (12 pc at the galactic center) along and 2.6' (8 pc) perpendicular to the plane. This distribution is remarkably similar to that of the dust emission longward of 50 microns in the same beamsize (see Becklin, Gatley and Werner[5]). Somewhat higher spatial resolution in OI will be required in order to check for a central "hole" in the emission similar to that seen by the latter authors. The distribution of OI is, however, quite different from that of the ionized gas. The free-free radio emission from Sgr A West[2,6] reveals a cluster of compact ionized clouds with an overall diameter of about 20" surrounded by a weaker "halo" about 1' across. The compact clouds were first detected by their 12.8 micron NeII emission.[8]

### 3. Kinematics

The OI lines are spectrally resolved at all positions, and the deconvolved linewidths range from 70 to 350 km-s$^{-1}$. The line center velocities change systematically along the galactic plane. South of the center, the velocity centroids are blueshifted with velocity $-20\pm 10$ km-s$^{-1}$. North of the center the line is redshifted with velocity $+115\pm 10$ km-s$^{-1}$. Above and below the plane the velocity is intermediate between these extremes. An extrapolation of the rotation curve of Sanders and Lowinger[9] inward to 1 pc predicts velocities of this magnitude. We feel that this lack of symmetry around zero velocity cannot be accounted for by

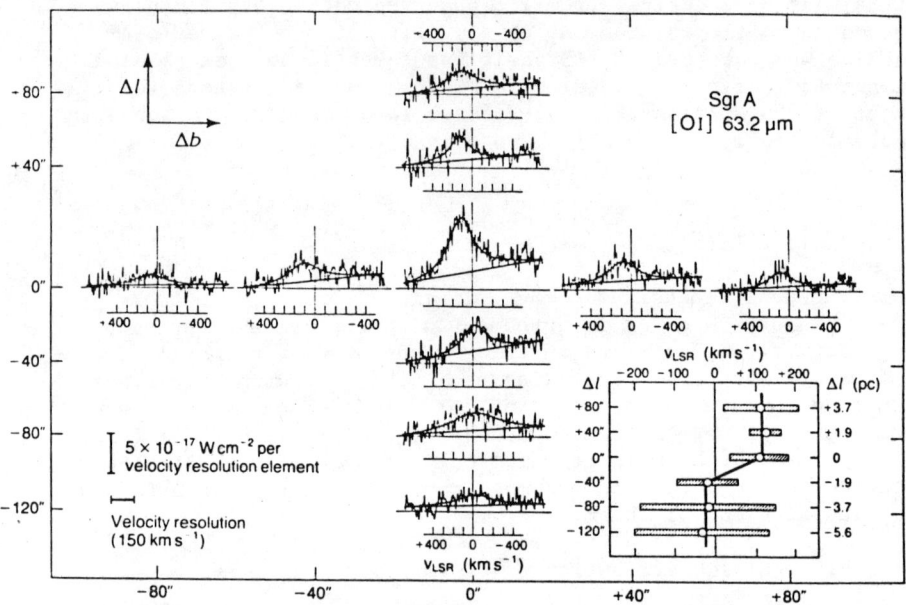

Figure 1. The observed spectra of the $3p_1-3p_2$ fine structure line of OI at 63.170 microns toward Sgr A in a 4' FWHM beam. The spectra are displayed as a function of offset from the map center in steps of 40". The map center is at (1950) $17^h\ 42^m\ 29^s\ -28°\ 59'$ which is about 15" north of the point radio source. Additional spectra around the periphery of this region do not show a line and are not presented here. The relative pointing accuracy is estimated to be about 10 arcseconds. The velocity resolution is 150 km-s$^{-1}$ and the velocity calibration is accurate to within 20 km-s$^{-1}$. Inset in the figure is a longitude-velocity diagram of the line centroid (open circles) and the deconvolved velocity widths (shaded bars). Note from this inset that the southern beam positions reveal gas at a more positive velocity than much of the gas in northern beam positions. These overlapping velocities are suggestive of non-circular motions in the galactic center region. All velocities are relative to the LSR.

errors in measurement or by uncertainty in the $H_2O$ line used for wavelength calibration (63.323 microns). The OI line is actually seen though the wing of this line in terrestrial absorption. While it is difficult to deconvolve the OI emission from this line, the result would be to increase the velocity asymmetry. This asymmetry may reflect either large-scale excitation or kinematic inhomogeneities in the emitting gas. The data indicate a rapid transition in the velocity over an angular distance of less than 40", centered on the position of the dynamical galactic center, which is actually about 15" south of our map center (0,0). The fact that the line widths are at least as large away from the center as at the center and the evidence for overlapping velocities (see Figure) around the center cannot be expained in terms of simple Keplerian rotation and suggests instead some kind of non-circular motion. The S/N in the OIII line was too small to yield information about the kinematics of the ionized gas in the beam.

4. Excitation of the OI Line

As pointed out by Lester et. al.[1] the OI emission can only arise from partially ionized or neutral gas, since the ionization potential of oxygen is only slightly greater than that of hydrogen. In predominantly neutral gas, the collisional excitation of the neutral oxygen triplet will be dominated by neutral hydrogen impact. The angular distribution of the OI line suggests that the emitting gas is coexistant with warm dust in the Sgr A West region. Becklin, Gatley and Werner[5] find that the dust has a peak brightness temperature of 50K at 63 microns. The observed peak brightness temperature in the line of about 130K is a lower limit to the excitation temperature of the OI and hence the kinetic temperature of the gas. Our data yield a luminosity of Sgr A West in the OI line of approximately $10^5$ $L_\odot$ (about 0.5% of the total continuum luminosity) assuming that the source is symmetrical around the observed positions. If we assume that the emitting gas is distributed uniformly throughout the region in a neutral medium with a density of $10^3$ $cm^{-3}$ (a value suggested by the far-infrared continuum observations) then a gas kinetic temperature of between 200K and 500K is required to produce the observed luminosity. A temperature of 130K can yield the observed luminosity if the region is clumped with a mean density of $>10^4$ $cm^{-3}$. A lower limit to the total mass of gas in the line emitting region is given by the high-density case, in which the OI line is collisionally deexcited[1], and with T>>228K. This lower limit is 300 $M_\odot$ (assuming a solar O/H ratio).

5. Density Structure of the OIII-Emitting Gas

The ratio of the 52 and 88 micron fine structure lines of OIII depends on the electron density of the emitting gas. Previously, Watson et. al.[2] detected the 52 micron $^3P_2-^3P_1$ line in Sgr A at an intensity of $5.2 \pm 2 \times 10^{-17}$ $W-cm^{-2}$. The resulting ratio ($0.13 \pm 0.05$) indicates an electron density of $>6 \times 10^3$ $cm^{-3}$ assuming electron impact excitation of $O^{++}$. This is significantly higher than the RMS electron density of 600 cm$^{-3}$ derived from the 1' halo of radio continuum emission in Sgr A West (see for example Rodriguez and Chaisson[10]). The difference between the densities inferred from the radio and line observations suggests that most of the ionized gas in Sgr A is in the form of dense clumps with electron densities $>>10^3$ cm$^{-3}$ and a small filling factor. This structure may thus be similar to that of the NeII emitting gas clumps seen in the central 20".[8]

## REFERENCES

1. Lester, D. F., Werner, M. W., Storey, J. W. V., Watson, D. M., and Townes, C. H. 1981 Ap.J. (Letters) 248, L109.

2. Storey, J. W. V., Watson, D. M., and Townes, C.H. 1980 International Journal of Infrared and Millimeter Waves 1, 15.

3. Gatley, I., Becklin, E. E., Werner, M. W., and Harper, D. A. 1978 Ap.J. 220, 822.

4. Dain, F. W., Gull, G. E., Melnick, G., Harwit, M. and Ward, D. B. 1978 Ap.J. (Letters) 221, L17.

5. Becklin, E. E., Gatley, I., and Werner, M. W. 1982 Ap.J.(in press).

6. Brown, R. L., Johnston, K., and Lo, K. Y. 1981 Ap.J. 250, 155.

7. Ekers, R. D., Goss, W. M., Schwarz, V. J., Downes, D., and Rogstad, D. H. 1975 Astr. Ap. 43, 159.

8. Lacy, J. H., Baas, F., Townes, C. H., and Geballe, T. R. 1979 Ap.J. (Letters) 227, L17.

9. Sanders, R. H. and Lowinger, T. 1972 Astr.J. 77, 292.

10. Watson, D. M., Storey, J. W. V., Townes, C. H., and Haller, E. E. 1980 Ap.J. (Letters) 241, L43.

11. Rodriguez, L. F., and Chaisson, E. J. 1979 Ap.J. 228, 734.

# THREE COMPACT SOURCES WITH UNUSUAL 2 TO 4 MICRON SPECTRA

S. P. Willner*
Center for Astrophysics and Space Sciences,
University of California, La Jolla, CA 92093
and
Harvard-Smithsonian Center for Astrophysics, Cambridge, MA 02138

J. L. Pipher*
Department of Physics and Astronomy,
University of Rochester, Rochester, NY 14627

## ABSTRACT

1.9 to 4.1-μm spectra of the compact infrared sources IRS 3, 4, and 19 are presented. IRS 3 lacks a strong ice absorption to accompany its enhanced silicate absorption. IRS 4 shows a 3.3-μm emission feature thought to arise from interfaces between molecular and ionized gas. IRS 19 shows an extremely deep 3.1-μm absorption superposed on the steam band that originates in a stellar atmosphere.

## INTRODUCTION

The region of the galactic center contains a number of compact infrared sources.[1,2] While the basic emission mechanism - thermal radiation from dust or stellar photospheres - is thought to be known for most of the sources on the basis of their broadband magnitudes,[2,3] relatively little detailed spectral information has been available for any of the sources. Other papers in this volume present new spectral observations primarily at wavelengths shorter than 2.8μm. This paper presents 1.9 to 4.1-μm spectra of three compact sources that have unusual spectral features.

## OBSERVATIONS

The observations were made with the CTIO 4-m telescope on 1980 June 29 for IRS 19 and July 3 for IRS 3 and 4. The spectrophotometer used circular variable filters as the wavelength selective element, and the spectral resolution ($\Delta\lambda/\lambda$) was 1.3% from 1.9 to 2.5μm and 1.1% from 2.8 to 4.1μm. A 2.5" beam size was used for IRS 3 and a 3.9" beam for the other sources. The sources were located by offsetting the telescope from IRS 7, then correcting the pointing by peaking up at the wavelength where the desired source was brightest with respect to nearby sources. The pointing corrections found by peaking were smaller than 1" in each coordinate. Once established, the telescope pointing was accurately maintained by reference to visible stars seen in a focal plane television camera. A chopping secondary

---

*Visiting astronomer at Cerro Tololo Interamerican Observatory, which is operated by the Association of Universities for Research in Astronomy, Inc. under contract with the National Science Foundation.

mirror was used to provide sky subtraction, and the chopper distance and telescope beam switch direction were chosen to minimize contamination of the observed spectra by other sources. We believe that contamination is negligible for these sources; the most serious potential problem is that at the shortest wavelengths IRS 3 may be slightly affected by the nearby IRS 7.

The observations were corrected for interstellar reddening on the assumption that the extinction is inversely proportional to wavelength and is 2.5 magnitudes at 2.2μm. Such a correction is in approximate agreement with other extinction estimates,[4] and in any case it is a smooth function of wavelength and thus cannot introduce spurious spectral features. The corrected spectra are presented in Figure 1, and the sources are discussed individually in the next sections.

## DISCUSSION - IRS 3

IRS 3 is a relatively red, thus dust-emitting, source from 2 to 5μm and has a stronger than normal silicate absorption.[3] It is spatially unresolved in a 1.5" beam.[2] It has been suggested that this source might be similar to compact infrared sources found in star formation regions[2,3] and referred to as "protostars."[5] Such a similarity would indicate ongoing star formation in the galactic center. Protostars usually, but not always, show an ice absorption feature near 3.1μm.[5]

Figure 1 shows that no strong ice absorption feature is present in IRS 3. There are, however, weak absorptions from 2.3 to 2.5μm and 2.8 to 3.6μm, the latter being largest near 2.9μm. These can be interpreted as CO and $H_2O$, respectively, in a heavily reddened late-type stellar atmosphere. (The 2.3 to 2.5μm feature is of low signal-to-noise and needs confirmation. It is not likely to be due to contamination from IRS 7, because the latter has a much bluer spectrum, and contamination would affect the shortest wavelength points the most.) On the other hand, the shape of the 2.9 to 3.6μm absorption is not really characteristic of either ice or steam; the former would be centered at 3.1μm, while the latter should not exhibit the observed turnup at 2.8μm. About all that can be said for certain is that the lack of strong ice absorption makes less likely - but cannot rule out - the protostar hypothesis.

The narrow absorption feature at 3.4 μm is clearly seen superposed on the much broader absorption duscussed above. The narrow interstellar feature[6,7] is no stronger in IRS 3 than in the other galactic center sources (e.g. IRS19), so the additional extinction necessary to explain the stronger silicate absorption in IRS 3 must be intrinsic to the source rather than interstellar.

## DISCUSSION - IRS 4

IRS 4 is spatially extended and has the same 5 to 20-μm energy distribution[8] as the low surface brightness background radiation that pervades the central arc-minute of the galactic center region.[2] IRS 4 also has a lower 10-μm surface brightness than most of the other

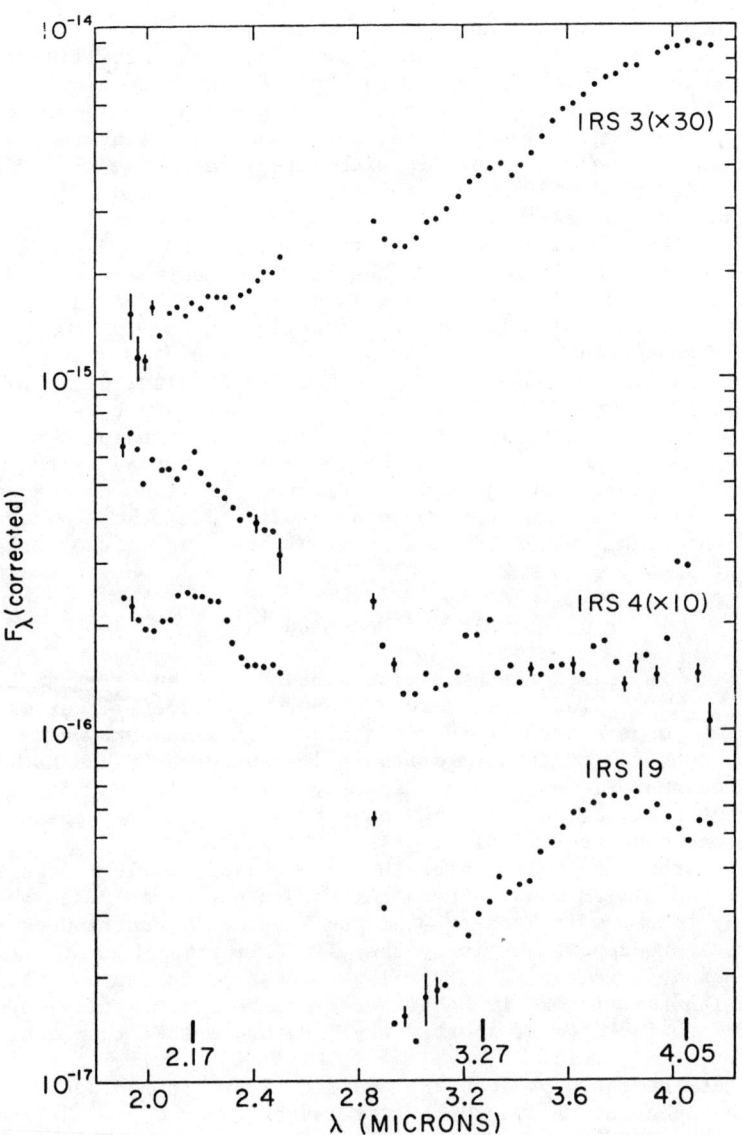

Fig. 1. Spectra of galactic center sources corrected for interstellar extinction. The correction is a factor of 10 at 2.2μm. Error bars are shown for all points whose statistical uncertainties exceed 5%.

compact sources and may represent a condensation in the extended background.[3] The red 5 to 20-μm energy distribution implies that dust emission is responsible for the bulk of the radiation.

The spectrum of IRS 4 is thus far unique among galactic center sources in showing a 3.3-μm emission feature. The feature is one of a set that is usually associated with interfaces between H II regions and molecular clouds.[6,9] Emission features at 6.2 and 7.7μm are part of the same set of features, and they should show even larger contrast than the 3.3-μm feature. Neither the 6.2 nor the 7.7-μm feature was seen in a 27" beam,[6] so the features must be unusual in the galactic center as a whole. IRS 4 is probably located in one of the few places in the galactic center where ionized and molecular gas are in direct contact.

The spectrum of IRS 4 in Figure 1 shows a relatively blue component emitting from 2.0 to 2.5μm. The surface brightness of this component is low, only twice as great as the average surface brightness in a 1.8' beam.[10] Much of the radiation observed is thus due to the extended stellar background.[10] Figure 1 shows no CO absorption near 2.4μm in this component; such absorption would be expected if most of the light is contributed by giant stars as in the nuclei of other galaxies.

## DISCUSSION - IRS 19

IRS 19 is relatively blue from 1.65 to 5μm, and its radiation is thought to be stellar. It is less than 2" in diameter, but whether it consists of a single or multiple star system is unknown.[4]

The spectrum of IRS 19 in Figure 1 shows deep CO and $H_2O$ absorption bands near 2.4 and 2.0μm respectively. The latter indicates a late spectral type, and probably only a single star is necessary to provide the observed emission.

The surprising thing about the spectrum of IRS 19 is the extreme depth of the 3.1-μm band. The steam band centered at 2.7μm and seen extending to 3.9μm in late-type stars[11] certainly contributes to the 3.1-μm feature depth, but in neither IRS 7 in the galactic center nor in nearby late-type giants[11] is the absorption depth at 3.1μm as much as 1/3 that in IRS 19. Furthermore, the steam optical depth continues to increase at shorter wavelengths, rather than decreasing at 2.8μm as in IRS 19.

Water ice could produce the additional absorption needed to explain the spectrum of IRS 19, but it is not clear where the ice could be located. The indicated ice optical depth is about 1.4, and in nearby molecular cloud sources, one would expect the corresponding silicate optical depth to be at least as high.[5] The silicate depth has not been measured for IRS 19, but its near infrared colors imply little or no additional additional extinction compared to other sources such as IRS 11 and 17, which show no ice absorption.

## CONCLUSION

Spectra show that infrared sources in the galactic center can have very different properties even if their broadband colors are similar. Both adequate spectral resolution and beam sizes small enough to separate the individual sources are needed to define the properties - and in some cases even the basic natures - of the compact sources.

The authors thank the staff of the Cerro Tololo Interamerican Observatory for their knowledgeable and enthusiastic help with the observations. This research was supported by the National Science Foundation.

## REFERENCES

1. E. E. Becklin and G. Neugebauer, Astrophys. J. Lett. <u>200</u>, L71 (1975).
2. G. H. Rieke, C. M. Telesco, and D. A. Harper, Astrophys. J. <u>220</u>, 556 (1978).
3. E. E. Becklin et al., Astrophys. J. <u>219</u>, 121 (1978).
4. E. E. Becklin et al., Astrophys. J. <u>220</u>, 831 (1978).
5. S. P. Willner et al., Astrophys. J. <u>253</u>, in press (1982).
6. S. P. Willner et al., Astrophys. J. Lett. <u>229</u>, L65 (1979).
7. D. T. Wickramasinghe and D. A. Allen, Nature <u>287</u>, 518 (1980).
8. G. H. Rieke and F. J. Low, Astrophys. J. <u>184</u>, 415 (1973).
9. D. K. Aitken et al., Astron. Astrophys. <u>76</u>, 60 (1979).
10. E. E. Becklin and G. Neugebauer, Astrophys. J. 151, 145 (1968).
11. K. M. Merrill and W. A. Stein, Pub. Astron. Soc. Pacific <u>88</u>, 285 (1976).

# TWO MICRON OBSERVATIONS of $^{12}$CO and $^{13}$CO in THE RED GIANT SOURCES IRS 7, IRS 12, and IRS 19

G. C. Augason
NASA, Ames Research Center, Moffett Field, California 94035

H. A. Smith
E. O. Hulburt Center for Space Research, Naval Research Laboratory, Washington, D. C. 20375

E. R. Wollman
Department of Physics and Astronomy, Bates College, Lewiston, Maine 04240

H. P. Larson
Lunar and Planetary Laboratory, University of Arizona, Tucson, Arizona 85721

and

H. R. Johnson
Department of Astronomy, Indiana University, Bloomington, Indiana 47405

We have obtained spectra from 1.2 to 2.47 microns of the Galactic Center sources IRS 7, 12 and 19 with resolutions of 2.4, 2.4, and 4.8 cm$^{-1}$ respectively. The observations were done with the U of A/LPL Fourier transform spectrometer at the CTIO 4 m telescope with a beam size of 3.8" (see Wollman et al., 1982). The spectra of these sources show the first overtone bandheads of $^{12}$CO and $^{13}$CO. The spectrum of IRS 7, the brightest of the sources, also shows photospheric absorptions of Ca, Na, and other common elements.

Spectrophotometric detection of CO in these sources first provided decisive evidence that the sources are cool giants or supergiants (Neugebauer et al. 1976). The signal-to-noise and resolution of the present spectra is adequate to measure the radial velocities from the positions of the CO bandheads. We find that all three sources are blueshifted: IRS 7 by $-170 \pm 50$ kms$^{-1}$, IRS 12 by $-70 \pm 75$ kms$^{-1}$, and IRS 19 by $-80 \pm 75$ kms$^{-1}$ (Wollman et al., 1982)

The bandheads of the first overtone of $^{13}$CO are clearly present in the spectra of IRS 7, 12, and 19. To determine the $^{12}$C/$^{13}$C ratios of IRS 7 and 12, synthetic spectra were computed using $^{12}$CO and $^{13}$CO molecular line parameters and 60 layer stellar atmosphere models. The atmospheric models were computed by the opacity sampling method which is described in Johnson et al., (1980). The CO line positions were computed using the constants of Mantz et al., (1975) and the strengths were computed using the semiempirical constants of Chackerian (1970). The micro and macroturbulent velocities were each set to 2 km/sec. Log g = 0 (supergiants) was used. (The bandhead spectra are not a strong function of log g). The elemental abundances were assumed to be solar with C/O = 0.6. The temperatures for IRS 7 and 12 were estimated, on the basis of the strengths of the V = 2-0, 3-1, and 4-2 $^{12}$CO bands, to be 3000 K, and 4000 K respectively. The total column density of CO was determined by varying the total CO in each model and then comparing the synthetic and observed spectra in the regions outside of the saturated bandheads. The total CO column densities for IRS 7 and IRS 12 were found to be $3 \pm 1 \times 10^{22}$ and $7 \pm 2 \times 10^{23}$ molecules/cm$^2$ respectively.

Fig. 1 shows the observed spectrum of IRS 7 between 4200 and 4300 cm$^{-1}$. The spectrum is unapodized and not interpolated between the computation points. The spectrum has many terrestrial features primarily due to $H_2O$ and $CH_4$. The strongest of these features are indicated in Fig. 1. Also, shown is the $3\sigma$ noise level. Superimposed on the observed spectrum are 3 spectra with $^{12}$C/$^{13}$C ratios of 90/1, 30/1, and 4/1. The spectra are shifted to compensate for the blue shift of IRS 7. The spectra are broken at 4240 cm$^{-1}$ to permit a better fit to the continuum which is depressed from 4240 to 4300 cm$^{-1}$ due to terrestrial features. The $^{13}$CO V = 3-1 bandhead at 4213 cm$^{-1}$ and the $^{13}$CO V = 2-0 bandhead at 4265 cm$^{-1}$ in Fig. 1 show that IRS 7 has a relatively high $^{12}$C/$^{13}$C ratio. Based on the $3\sigma$ noise level for IRS 7, $^{12}$C/$^{13}$C > 30. Using the same method of analysis for IRS 12, $^{12}$C/$^{13}$C > 30 also. The $^{13}$CO band heads are detected in IRS 19. However, because of the reduced signal-to-noise and resolution, no lower limit for $^{12}$C/$^{13}$C was established for IRS 19. The $^{12}$C/$^{13}$C limits presented here are affected by terrestrial atmospheric features and some uncertainties in the stellar atmospheric models. Further work to overcome these problems is continuing and will be reported in a future paper.

Chackerian, C. 1970, J. Quant. Spect. Rad. Trans., 10, 271.

Wollman, E. R., Smith, H. A., and Larson, H. P. 1982, ApJ., (Submitted).

Johnson, H. R., Bernat, A. P., and Krupp, B. M. 1980, ApJ. Suppl., 42, 50.

Mantz, A. W., Maillard, J. P., Roh, W. B., and Rao, K. N. 1975, J. Mol. Spect., 57, 155.

Neugebauer, G., Becklin, E. E., Beckwith, S., Matthews, K., and Wynn-Williams C. G. 1976, ApJ., 205, L139.

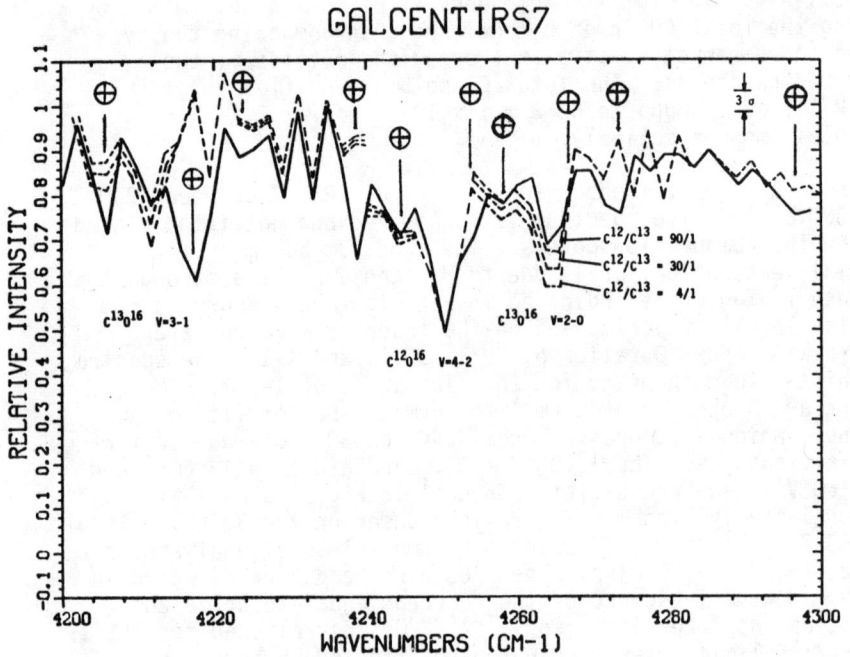

Figure 1. Three synthetic stellar spectra (dashed lines) with $^{12}C/^{13}C$ = 90/1, 30/1, and 4/1 superimposed on the observed spectrum of IRS 7 (solid line). The major atmospheric features are indicated.

# MAPPING AND IMAGING OF THE GALACTIC CENTRE IN THE NEAR INFRARED.

J.W.V. Storey
Anglo-Australian Observatory, P.O. Box 296, Epping, N.S.W. 2121
Australia.

## ABSTRACT

A cooled CCD has been used to produce an image of the Galactic centre at an effective wavelength of 0.9 microns. The image shows two unresolved point sources separated by 3 arcseconds along the Galactic plane. These sources are very close to the compact non-thermal radio source, and the 2.2 micron source IRS-16. While it is possible that the new sources are a pair of highly reddened individual stars or compact clusters, it appears more likely that they are two compact HII regions seen via line emission. This conclusion is supported by a preliminary map of the region in Brackett $\gamma$. The 1.2 micron continuum map shows an enhancement near one of the sources, but the other source is not apparent.

## OBSERVATIONS

The CCD image[1] (figure 1) was taken at the prime focus of the Anglo-Australian Telescope on 8 September 1981, during conditions of sub-arcsecond seeing. Built for the AAO by the Royal Greenwich Observatory, the CCD camera uses a thinned, back-illuminated RCA SID 53612 buried-channel CCD with 320 x 512 pixels. The resulting image scale at the f/3.3 focus is 0.50 arcseconds/pixel. The exposure time for figure 1 was 10 minutes through an RG830 (8300 A long-pass) filter. This particular CCD system has a quantum efficiency which falls linearly from about 60% at about 8000 A to zero at one micron. Thus, together with the RG830 filter, an effective bandpass of about 1000 A centred on 9000 A is defined.

By comparing figure 1 with a IV N Schmidt plate (courtesy of the UK Schmidt Telescope Unit) and with an R-band exposure taken with the CCD, all the stellar images can be identified with previously known stars except for the two images marked CCD1 and CCD2. These two new sources must therefore be very red, and the most probable explanation is that they are suffering the high extinction expected for sources on the Galactic plane at the distance of the Galactic centre.

Positions for the new sources were obtained by measurement relative to two bright field stars, visible on figure 1. These stars were in turn identified on the Palomar Sky Survey E plate, and their positions determined relative to fifteen nearby SAO stars by measurement of the plate on a microdensitometer, followed by a 6-parameter least-squares fit. The northernmost of the two bright visible stars on figure 1 (labelled "A") is the visible star identified on the 2.2$\mu$m map of Becklin and Neugebauer[2]; the position determined by those authors is in good agreement with the present value.

The intensities of the new Galactic centre sources were

determined by observing the white dwarf LDS749B with the CCD / filter combination. The resulting magnitudes for CCD1 and CCD2 are 18.9 ± 0.3 and 19.1 ± 0.3 respectively.

Partial maps of the environs of the two images were made in the continuum at 1.2 $\mu$m and in the Brackett $\gamma$ emission line by Allen, Carter and Malin[3] using the infrared photometer-spectrometer on the AAT. They used a 2 arcsec beam and explored an area 5-6 arcsec

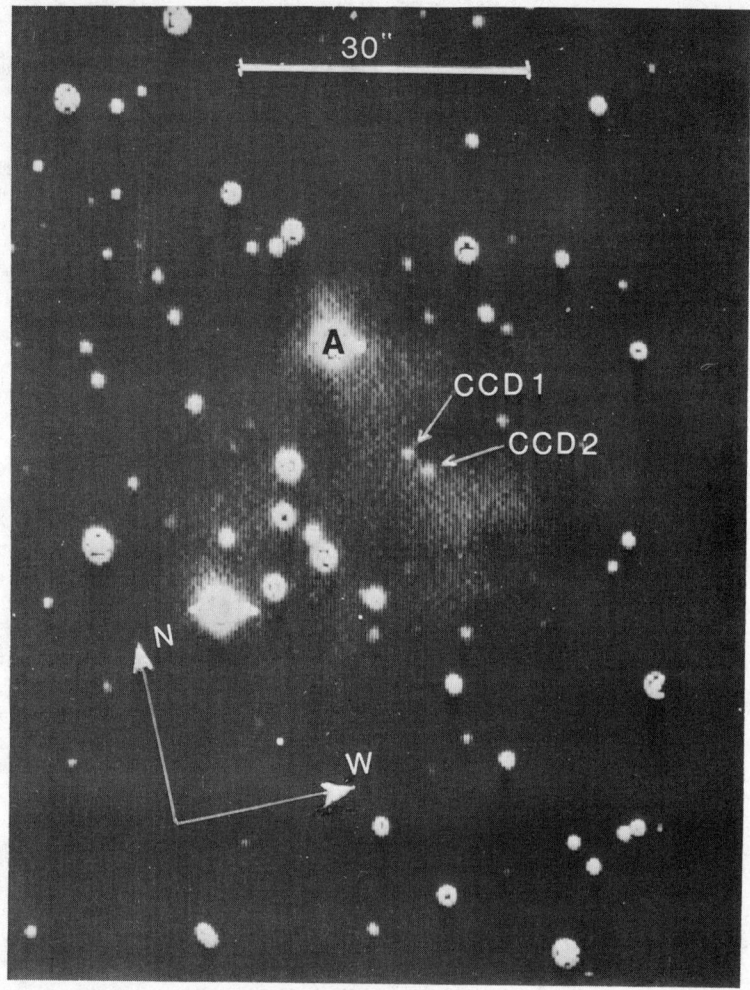

Figure 1. CCD photograph of the Galactic centre at an effective wavelength of 0.9 $\mu$m. All the stellar images are also seen on a IV N plate, with the exception of the two faint objects (CCD1 and CCD2) indicated. The faint nebulosity may not to be real.

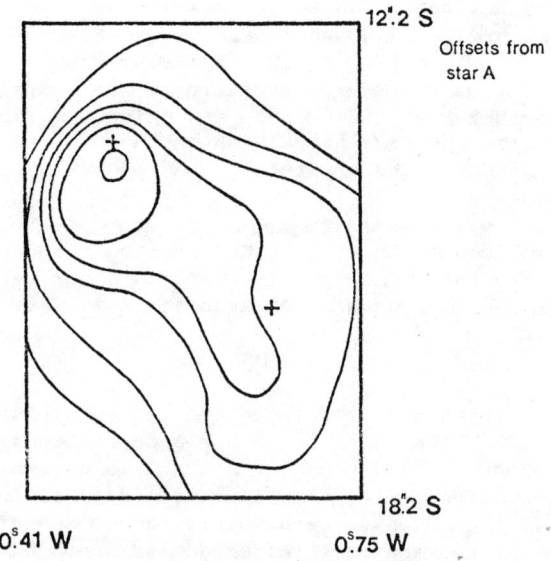

Figure 2. Partial map of the region around the CCD images at 1.2 μm. The crosses mark the positions of the two CCD sources.

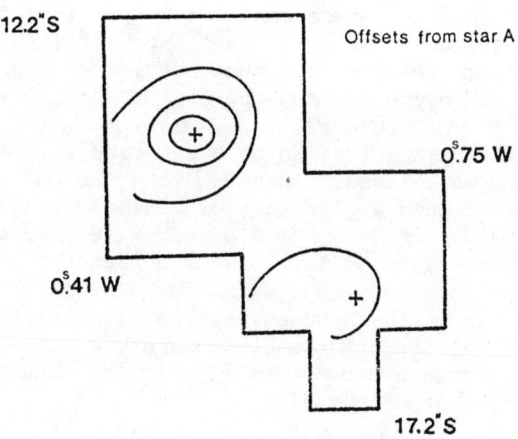

Figure 3. Partial map of the region around the CCD images in Brackett γ. The crosses mark the positions of the two CCD sources.

on a side. Sky reference was taken from a region 20 arcsec south of the Galactic centre which appears dark on the 2.2μm continuum map by Allen, Hyland and Jones (reported at this workshop by Gatley). In the 1.2 μm continuum (figure 2) an elongated bar envelopes the two CCD objects, brighter at its NE end, and quite similar to the 2.2 μm continuum map. The point sources seen on the CCD map are not themselves prominent. The entire feature appears as IRS-16 on Becklin and Neugebauer's map[1].

In Brackett γ two emission knots are seen, coincident to within less than 1 arcsecond with the two CCD sources (figure 3). No ridge of emission links them, and the NE object is about twice as intense as the other. The sources are believed to be isolated, though the map is not complete.

## DISCUSSION

If the flux in each of the sources detected by the CCD is due to stellar continuum, then each source has a de-reddened absolute magnitude of about -7.7. Because each image is unresolved it would be reasonable to ascribe all of this flux to a pair of individual stars. Each star would then need to be a fairly early supergiant. These CCD results represent the shortest infrared wavelength at which the Galactic centre has been observed, and hence it is tempting to suggest that the sources detected by the CCD are hot, early stars which provide the ionising flux for the region. Although the total flux required to power the "ridge" HII regions could be provided by a single O5 star, Lacy et al[4] have shown that this flux must be unusually soft, i.e. $T_{eff} \approx 35,000$ K (see also Watson et al[5]), corresponding to something closer to an O9 star. It thus appears unlikely that a single pair of stars could be responsible for all the ionisation. One cannot, however, rule out the possibility that the stars seen by the CCD are members of a somewhat larger cluster, whose overall ionising output powers the region.

From the infrared observations[2] we can derive a magnitude at J for CCD1 of 14.6, while at K the value is 9.2. From this J-K colour of 5.4 an $A_V$ of 34 can be derived. Assuming that the CCD images are due to fairly hot stars, then application of a simple reddening law results in a predicted magnitude of 21.3 at 9000A. Comparison of this figure with the observed value of 19.1 suggests that the CCD images must arise at least in part from line emission. The major uncertainty, however, is in the bandpass of the CCD / filter combination, and hence of the effective wavelength of the observations. This uncertainty prevents any firm conclusions from being drawn on the basis of colour alone.

If the powerful 2.2 μm source IRS-7 is a late M supergiant as is normally assumed, its apparent magnitude at 0.9 μm would also be around 20. This is not greatly weaker than the sources detected by the CCD, and suggests that a somewhat deeper exposure would make this source detectable as well.

The close correspondence between the Brackett γ map and the CCD picture strongly suggests that the flux at 0.9 μm is largely due to line emission, and hence that CCD1 and CCD2 are compact H II regions rather than stars. The most plausible lines are the [S III]

$3p^2$ $^1D \rightarrow {}^3P$, $J = 2 \rightarrow 1$ and $J = 2 \rightarrow 2$ transitions at 9069 and 9532 Å respectively, with some additional contribution from the Paschen series of hydrogen. The helium I 10830 Å line would probably be too far down on the CCD response to be detectable.

Unfortunately, the best available VLA map[6], taken at 5GHz, has insufficient resolution to allow proper identification of the CCD images. The VLA map does, however, make it fairly clear that the powerful non-thermal radio source at the Galactic centre consists of a single component only. Unfortunately, the different astrometric reference frames used at radio and optical wavelengths do not allow registration of the two maps to better than about 1 arcsecond. Thus one cannot rule out completely the possibilty that the non-thermal radio source lies between the CCD images, although this appears unlikely.

If the CCD images are in fact compact H II regions, then because of their apparent similarity and their proximity to the the powerful non-thermal radio source it is possible that they are actually ionised by this source. This does not at present tell us anything new about the nature of the central source; however it is clear that further studies of the ionisation state and dynamics of the CCD sources will prove to be particularly informative.

## FUTURE WORK

At present it is not possible to determine the nature of the CCD sources with any certainty. Within the next few months, work at the AAT will be directed towards:

1) Obtaining a more complete Brackett $\gamma$ map of the region at high spatial resolution.
2) Better determination of the infrared colours of the CCD objects.
3) Obtaining spectra of the two objects over the range 0.9 to 1.0 $\mu$m. Sufficient reslution should be achieved to explore the dynamics of the two sources, and hence to place constraints on the mass density in the vicinity. (Assuming of course that line emission is present.)
4) Imaging of the central few parsecs both through broadband far-red filters and through a [S III] 9532 Å interference filter.

It is also important to obtain radio maps at as high a spatial resolution as can be achieved by the VLA. This needs to be done in conjunction with improvements to the registration of the optical and radio frames of reference. Other relevant information to be sought includes [Ne II] and [S IV] spectra at 10 $\mu$m and high spectral resolution mapping of the area in Brackett $\gamma$.

## ACKNOWLEDGEMENTS

The CCD picture discussed in this presentation is a result of work done in colaboration with J.O. Straede, J.V. Wall, P.R. Jorden and D.J. Thorne. We wish to thank the Director of the AAO for allocating time to this project during the commissioning of the CCD. Thanks are also due to D.A. Allen, D. Carter and D.F. Malin for the infrared observations.

Table I Positions of Galactic Centre Sources.

| Object | R.A. (1950.0) | Dec (1950.0) |
|--------|---------------|--------------|
| CCD1   | 17  42  29.47 | -28  59  17.4 |
| CCD2   | 17  42  29.53 | -28  59  15.0 |

## REFERENCES

1. Storey, J.W.V., Straede, J.O., Jorden, P.R., Thorne, D.J. and Wall, J.V., submitted to Nature, 1982.
2. Becklin, E.E. and Neugebauer, G. Astrophys. J. Lett. $\underline{200}$, L71 (1975).
3. Allen, D.A., Carter, D. and Malin, D.F., Private communication (1981).
4. Lacy, J.H., Townes, C.H., Geballe, T.R. and Hollenbach, D.J. Astrophys. J. $\underline{241}$ 132 (1980).
5. Watson, D.M., Storey, J.W.V., Townes, C.H. and Haller, E.E. Astrophys. J. Lett. $\underline{241}$ L43 (1980).
6. Brown, R.L. Johnston, K.J. and Lo K.Y. Astrophys. J. $\underline{250}$ 155 (1981).

## TWO COLOR CCD OBSERVATIONS OF THE GALACTIC CENTER

J. A. Biretta, K. Y. Lo, and P. J. Young[*]
California Institute of Technology, Pasadena, Ca. 91125

### ABSTRACT

We present intermediate band photometric and astrometric observations at 8000Å and 9200Å of the galactic center region. Two unresolved objects are found near the positions of the compact non-thermal radio source in Sgr A and the infrared source IRS16. If either of the objects we find is at the distance of the galactic center, its emission must be non-stellar in nature. These objects may also be foreground stars. Spectroscopic observations are needed to determine unambiguously the nature of these objects.

### INTRODUCTION

The compact non-thermal radio source in the galactic center[1,2] is nearly coincident with the 2 micron source IRS16, the best candidate for the central star cluster[3]. To determine the mass of the underlying energy source of the radio emission, spectroscopic observations of the velocity dispersion of stars surrounding the radio source are necessary. Because the galactic center is highly obscured, identification of either the central star cluster or of the compact radio source must be made in the near infrared.

With the advent of charged-coupled-device (CCD) detectors of high quantum efficiency near 1 micron, accurate photometric and high resolution spectroscopic observations become feasible. Previous efforts to identify the compact radio source with photographic methods[4] led to suggestive results. Here we report 2-color photometric and astrometric CCD observations near 1 micron of the galactic center.

### OBSERVATIONS AND REDUCTIONS

Pictures of the galactic center region were obtained on 1980 June 17. Exposures were made through an i filter[5] and a z filter[6]. These filters have effective center wavelengths of 8000Å and 9200Å respectively. A 500x500 pixel CCD detector was used on the Prime Focus Universal Extragalactic Instrument camera[7] on the 5-m Hale telescope. We show a small region of the z band picture in Figure 1. The objects near Sgr A and IRS16 are labeled A and B. The stars labeled 1 and 2 were used as reference stars for astrometric measurements.

Astrometric measurements were made with the z band data. Image centers were found by fitting model star profiles to the images. The right ascension and declination offsets from the bright reference

---

[*]deceased September 1981

stars to the objects were then derived.

Photometric measurements were obtained for both A and B and for all objects in a 50" by 50" box centered between A and B (Figure 1). Photometric calibration was done using standard stars of the uvgriz system[5,8,9].

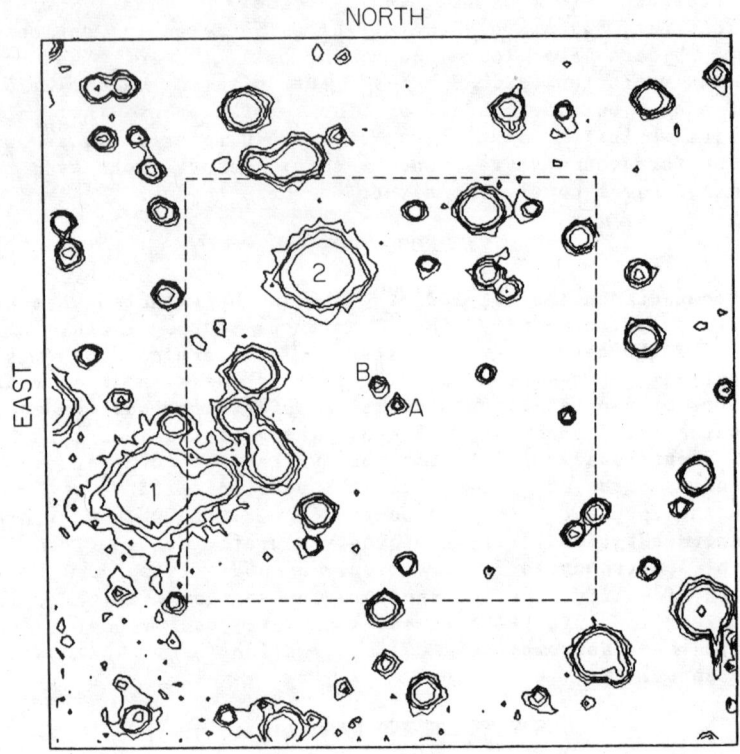

Figure 1. Averaged z band picture showing objects A and B. Astrometry reference stars 1 and 2 are also labeled. The 50" by 50" box shows the region where photometry was done on field objects.

## ASTROMETRIC RESULTS

Objects A and B are both unresolved by our 1.7" (FWHM) beam. Table I shows the measured position offsets from the reference stars to objects A and B. Quoted errors are 1 σ.

Given the coordinates of the reference stars[10] absolute positions of the objects can be obtained. These are presented in Table II along with the positions of the compact non-thermal radio source[11,12] and of IRS16[13,14]. Systematic errors limit the accuracy of positions presented in Table 2.

Table I. Position offsets from reference stars to objects.

| Reference Star | Object | Offset (1950.0) α | δ |
|---|---|---|---|
| 1 | A | -29.76±.08" | 9.55±.06" |
|   | B | -27.65±.08 | 11.84±.06 |
| 2 | A | - 8.71±.06 | -16.12±.07 |
|   | B | - 6.61±.06 | -13.83±.07 |

Nominally, object A, the compact radio source, and IRS16 are coincident to within the error bars. Given the uncertainties in the systematic errors, identification of the objects detected at the various wavelength bands must be confirmed by better astrometric observations.

Table II. Positions of Sources.

| Object | α (1950.0) | δ (1950.0) | Reference | Object |
|---|---|---|---|---|
| A | $17^h42^m29.30^s$±.07$^s$ | -28°59'18.1"±1.1" | Star 2 | (a) |
| B | 17 42 29.46 ±.07 | -28 59 15.8 ±1.1 | Star 2 | (a) |
| Compact | 17 42 29.335±.008 | -28 59 18.6 ±0.24 | 1730-130 | (b) |
| Radio Source | 17 42 29.35 ±.02 | -28 59 18.2 ±0.5 | 1748-253 | (c) |
| IRS16 | 17 42 29.3 ±.1 | -28 59 18.0 ±1.5 | | (d) |
|  | 17 42 29.35 ±.1 | -28 59 20.0 ±1.5 | | (e) |

(a) Assumed position of star 2 (1950.0):
   α=$17^h42^m29.96^s$±.07$^s$  δ=-28°59'02.0"±1.0"
(b) Brown, Johnston, and Lo (1981).
(c) Brown and Lo (1981).
(d) Becklin and Neugebauer (1975).
(e) Neugebauer, Matthews, and Soifer (1982).

## PHOTOMETRIC RESULTS

i magnitudes and i-z colors for objects A and B are presented in Table III. Quoted errors are 1 σ.

To see if either object A or B was unique we measured i magnitudes and i-z colors on all objects with 15<z<21 in a 50" by 50" box centered between A and B. The boundaries of this box are shown in Figure 1. Apparent i-z colors and z magnitudes for all these objects are presented in Figure 2.

Table III. i and z magnitudes of objects.

| Object | i | z | i-z |
|---|---|---|---|
| A | 20.80 ± .11 | 19.40 ± .07 | 1.40 ± .16 |
| B | 20.45 ± .11 | 19.32 ± .07 | 1.13 ± .16 |

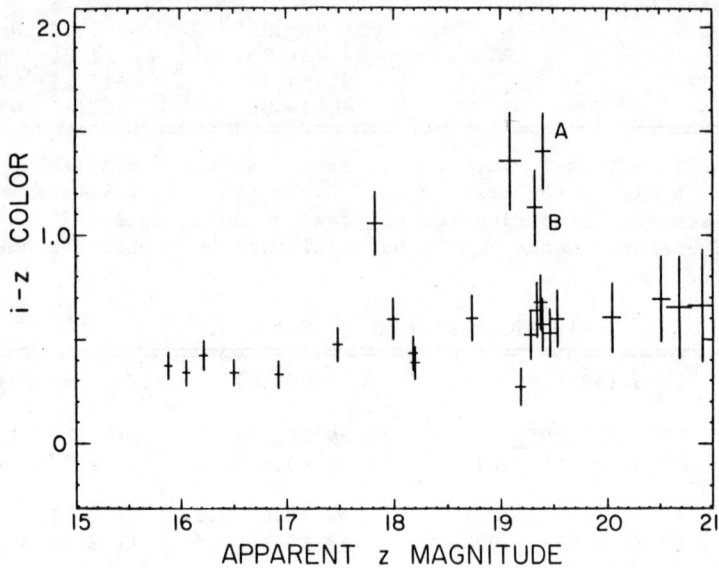

Figure 2. Plot of i-z color vs. apparent z magnitude for all objects with 15<z<21 in a 50" by 50" box centered between objects A and B. Error bars are one σ.

The objects may be divided into two groups on the basis of i-z color: (1) Objects with 0.2<i-z<0.7. These may be understood as a population of K and M dwarf stars at a mean distance of about 1 kpc. (2) A group of red objects with i-z$\geq$1.1. This group contains both A and B as well as other objects 20" or more distant from A and B. From this it is apparent that A and B are redder than typical objects in this field. However there is at least one other object in the field with a similar i-z color and z magnitude.

Of the objects found by Becklin and Neugebauer[13] at 2 microns, only IRS1, IRS9, IRS10, IRS11, and IRS16 may be coincident with an object we observed at 9200Å. The other 2 micron objects must be fainter than z=21.5.

## DISCUSSION

The effects of reddening must be considered when interpreting magnitudes and colors of objects toward the galactic center. Using van de Hulst's theoretical curve no. 15[15,16] and $A_K$=2.7$\pm$.5 to the galactic center[3] we find $A_z/E_{i-z}$=3.8 and $A_z$=12.3$\pm$2.5 to the galactic center. We will assume that extinction (in magnitudes) is proportional to distance. The corresponding dereddening track for object A is shown in Figure 3. If object A were at the galactic center, then $M_z$=-7.8$\pm$2.5 and i-z=-1.8$\pm$.6. The error bars on this point are due to the uncertainty in the reddening to the galactic center[3]. The track for object B is very similar. Also plotted in Figure 3 are absolute z magnitudes and i-z colors for common stellar types.

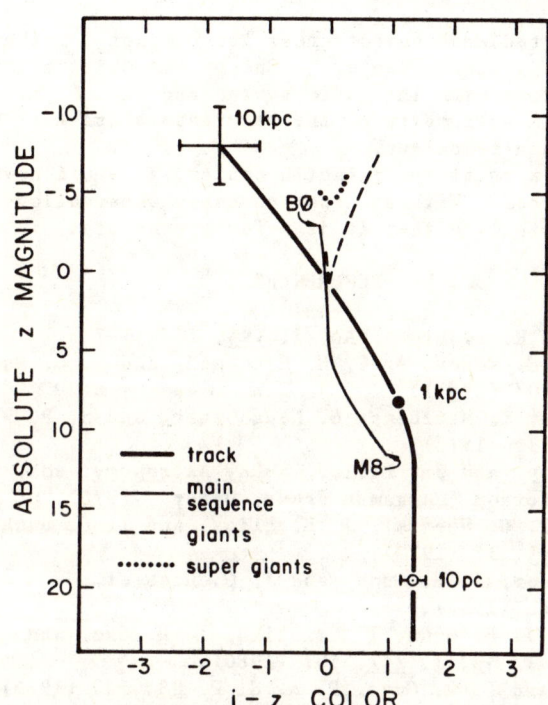

Figure 3. Plot of intrinsic absolute z magnitude ($M_z$) vs. intrinsic i-z color. The heavy line indicates the dereddening track for object A. This is the locus of possible intrinsic $M_z$ and i-z of object A depending on its distance from the sun. The track for object B is very similar. The stellar main sequence, giant branch, and super giant branch are indicated.

From Figure 3 we see that if object A were at the distance of the galactic center it would be significantly bluer than any common stellar type. Hence the emission would most likely be non-stellar in origin. This i-z color corresponds to a spectral index of α~14 where $S \propto \nu^{\alpha}$.

Objects A and B could also be foreground stars. Figure 3 shows that the dereddening tracks for both cross the main sequence and the giant branch. Thus, either or both objects could be giants or main sequence stars with appreciable reddening. This is perhaps the simplest interpretation of the observed magnitudes and colors. All of these interpretations, however, are critically dependent on the galactic center extinction in the i and z bands. These have been derived from the 2 micron extinction which is rather uncertain[3] and from van de Hulst's theoretical curve no. 15 [15,16] which may differ from the actual interstellar extinction curve for the galactic center.

## SUMMARY

We find two reddened objects near the compact non-thermal radio source in the galactic center. One of the objects (object A) is nominally coincident with the radio source and the 2 micron source IRS16. However, astrometry of the reference stars must be improved to confirm any identification.

If object A were at the galactic center, it would have to be a non-stellar source. Without spectroscopic observations, we cannot rule out the possibility that it is a foreground star.

## REFERENCES

1. B. Balick and R. L. Brown, Ap. J. 194, 265 (1974).
2. K. Y. Lo, M. H. Cohen, A. C. S. Readhead, and D. C. Backer, Ap. J. 249, 504 (1981).
3. E. E. Becklin, K. Matthews, G. Neugebauer, and S. P. Willner, Ap. J. 220, 831 (1978).
4. J. E. Grindlay, and Wm. Liller, X-Ray Astronomy, eds. W. A. Baity and L. E. Peterson (Pergamon Press, Oxford, 1978), p. 83.
5. R. A. Wade, J. G. Hoessel, J. H. Elias, and J. P. Huchra, P. A. S. P. 91, 35 (1979).
6. D. P. Schneider, J. E. Gunn, and J. G. Hoessel, in preparation (1982).
7. P. J. Young, J. E. Gunn, J. Kristian, J. B. Oke, and J. A. Westphal, Ap. J. 241, 507 (1980).
8. T. X. Thuan and J. E. Gunn, P. A. S. P. 88, 543 (1976).
9. D. P. Schneider, private communication (1981).
10. J. Lacy, private communication (1981).
11. R. L. Brown, K. J. Johnston, and K. Y. Lo, Ap. J. 250, 155 (1981).
12. R. L. Brown and K. Y. Lo, Ap. J., in press (1981).
13. E. E. Becklin and G. Neugebauer, Ap. J. Lett. 200, L71 (1975).
14. G. Neugebauer, K. Matthews, and B. T. Soifer, in preparation (1982).
15. H. C. van de Hulst, Recherches Astr. Obs. d'Utrecht 11, Part 1.
16. H. L. Johnson, Nebulae and Interstellar Matter, eds. B. M. Middlehurst and L. H. Aller (University of Chicago Press, Chicago, 1968), p. 167.

Discovery of Three Near Infrared Objects
in CCD Images of the Galactic Center

G.R. Ricker[*], M.W. Bautz[*], D. L. DePoy[*], and S.S. Meyer
Department of Physics and Center for Space Research
Massachusetts Institute of Technology
Cambridge, MA  02139

ABSTRACT

At visible wavelengths, the light of the Galactic Center is extinguished by a factor of $\sim 10^{12}$ due to absorbing matter along the 10 kpc path to the Sun. Thus, direct visible observations are not possible. However, at "far red" wavelengths ($\lambda > 7000 Å$), the extinction declines dramatically, dropping to $\sim 10^5$ near a wavelength of $0.9 \mu m$. Recently, sensitive charge-coupled devices (CCDs) have been developed with high quantum efficiency in the $0.9 \mu m$-$1.0 \mu m$ range. On 1981 September 4/5 and 5/6, we used a CCD detector system (the "MASCOT") at the Cerro Tololo Inter-American Observatory, to detect 3 extremely red objects in the direction of the Galactic Center which were not previously known. All three objects are <10" from the location of Sgr A West. If these objects do indeed lie at a distance of 10 kpc, then they are located within the innermost 1 pc of the Galaxy. One of these objects, designated NIR 1, lies within the extended $2.2 \mu m$ infrared source IRS 16, which is thought to contain the dynamic center of the Galaxy. Two interpretations which we consider for these three objects (NIR 1, NIR 2, and NIR 3) are that they are either tightly knit clusters of K5-M0 giant stars, or that they are extremely compact H II regions seen in the light of forbidden, doubly-ionized sulfur. Further observational tests of these two hypotheses are suggested.

INTRODUCTION

Investigations of the Galactic Center region with increasing spatial and spectral resolution over the past decade[1-6] have revealed an intriguing heirarchy of structures at wavelengths extending from the radio, through the infrared, and into the X- and $\gamma$-ray energy ranges. Notably absent has been information from the ultraviolet and visible regimes, because of the heavy extinction

---

[*]Visiting Astronomer, Cerro Tololo Interamerican Observatory, La Serena, Chile, which is operated by the Association of Universities for Research in Astronomy, Inc., under contract with the National Science Foundation.

toward the galactic nucleus. However, with the recent maturation of charge-coupled device (CCD) detector technologies, it has become possible to construct highly efficient imaging sensors which operate in the "far red" (~0.7-1.0μm) spectral region, straddling the border between the visible and IR regimes. Since the obscuration due to interstellar gas and dust falls very rapidly toward longer wavelengths in the 0.5-1.0μm range[7,8] there is good reason to believe that a direct image* of the Galactic Center might be obtained in the 0.8-1.0μm range with a suitable CCD system.

## INSTRUMENTATION

The instrument used to obtain the CCD images of the Galactic Center was the MASCOT (MIT Astronomical Spectrometer Camera for Optical Telescopes)[10]. The MASCOT utilizes two independent CCDs. One serves as the detector for a long slit grating spectrometer, while the other functions as an area imager. Both "channels" of the instrument are optimized for use at Cassegrain foci on telescopes as fast as f/7.5. The two CCDs are Texas Instruments 490x328 "virtual phase" arrays.[11,12] For the observations discussed here, the ability of the MASCOT to accomodate a variety of internal transfer lenses in the direct imaging "channel" proved to be extremely useful, permitting an optimum match of the pixel size to the seeing disk. Additional optical details of the instrument are given in Meyer and Ricker[13], while Dewey and Ricker[14] have described the methods used in the MASCOT system for data acquisition and on line image processing, as well as instrument control. Among the notable features of the MASCOT are its high throughput (~15 percent referred to the top of the atmosphere near 6000A, including telescope light loss , for both the area imager and for the spectrometer) and relative immunity to interference fringe difficulties (the virtual phase CCDs are front-side illuminated, epitaxial devices). Additional performance details and sample images of faint objects have been presented by Ricker et al.[10]

## OBSERVATIONS AND RESULTS

During the week 1981 September 1-7 the MASCOT was mounted at the Cassegrain focus of the CTIO 1.5m telescope. During this period, ~20 hours of observations of the Galactic Center region were obtained. The MASCOT was equipped with a 240A FWHM interference filter centered at 9150A. Transfer optics were installed which resulted in a CCD field of view of 5!3 NS x 7!1 EW, with an

---

*Grindlay and Liller[9] reported a possible detection of the Galactic nucleus itself in observations made in 1978 at 0.8μm. These results were based on "plate stacking", and have not yet been confirmed.

image scale of 1".3/pixel.  On two nights (4/5 September and 5/6 September), seeing of 1".0-1".5 FWHM was experienced during integration times of 1-2 hours.  Figure 1 represents the central portion of one such image, which was centered near the coordinates of the nonthermal source in Sgr A West.[4]  Corrections for pixel-to--pixel variations in quantum efficiency have been made using the technique described by Ricker et al.[10] and the processed image has been reduced to an isophotal contour plot to facilitate quantitative analysis.  In the image, there appear three objects which are not evident on copies of either the POSS blue or red plates, the SRC I survey plate, or the IV-N plate of the Galactic Center published by Liller and Alcaino.[9]  Thus, it would appear that these objects are extremely red, and are quite possibly near the inner edge of the Sagittarius arm[15] or even possibly in the vicinity of the Galactic Center itself.  The reality of these objects is assured by their appearance on the 4 different raw images which we have analyzed thus far, one of which was taken on 4/5 September and three of which were taken on 5/6 September.  The signal-to-noise ratio of each object exceeds 3:1 on all of the images.  Furthermore, the field was repositioned at different locations on the CCD between successive exposures.

Coordinates for these sources were determined relative to stars "A" (Figure 1 and Table 1) and "B" (Table 1), which by convention have been used for this purpose in IR studies of the Galactic Center region.[1,16]  The coordinates for these three objects (which we have designated NIR 1, NIR 2, and NIR 3) are given in Table 1, along with those of nearby 2.2μm and radio objects.  The coordinates we adopted for Stars A and B are also given in Table 1, along with the magnitudes and uncertainties (systematics included) which we have estimated for NIR 1, NIR 2, and NIR 3.  The magnitudes given are in the AB system of Oke[17], and were calibrated with respect to two IIDS standards (Hiltner 600 and Feige 15).[18]

The flux detected from each of these sources is $\sim 1 \times 10^{-14}$ ergs cm$^{-2}$ s$^{-1}$.  For an extinction correction of 13 magnitudes at 0.6μm (estimated from Becklin et al.[8], corresponding to 30 magnitudes of visual extinction), this implies a de-reddened flux of $\sim 1.5 \times 10^{-9}$ ergs cm$^{-2}$ s$^{-1}$.  The map of the Galactic Center region constructed by Becklin and Neugebauer[2] at 2.2μm is the one closest in wavelength and resolution to our image.  We have reproduced their map, with our three sources superimposed, as Figure 2.  Our three sources appear close to, but not coincident with, three of the 2.2μm sources:  NIR 1 is 2".4 from IRS 16, NIR 2 is 3".3 from IRS 1, and NIR 3 is 2".4 from IRS 9.  In a later paper, Becklin et al.[19] point out that the positions of the various objects in their map are uncertain by about ±2" in both coordinates.  For the case of IRS 16, there is the additional consideration that it is extended by ~2" at 2.2μm.[19]  Thus, a possible correspondence between the

Table 1: Positions and Magnitudes of Objects Observed
in the Galactic Center Region

| Object | R.A. (1950) | Dec (1950) | $m^*_{9150}$ |
|---|---|---|---|
| NIR 1[**] | $17^h42^m29^s.22$ | $-28°59'\ 15".8$ | $18.7 \pm 0.4$ |
| IRS 16[†] | $29^s.3$ | $18"$ | ---- |
| Sgr A West[††] (nonthermal) | $29^s.335$ $\pm0^s.008$ | $18".60$ $\pm0".24$ | ---- |
| NIR 2[**] | $29^s.42$ | $14".6$ | $19.5 \pm 0.4$ |
| IRS 1[†] | $29^s.6$ | $17"$ | ---- |
| NIR 3[**] | $29^s.45$ | $24".4$ | $19.7 \pm 0.4$ |
| IRS 9[†] | $29^s.6$ | $23"$ | ---- |
| Star A[#] | $29^s.95$ | $01".8$ | ---- |
| Star B[#] | $31^s.55$ | $27".5$ | ---- |

[*] $m_{9150} = -2.5 \log f_\nu -48.6$; $f_\nu$ in ergs cm$^{-2}$ s$^{-1}$ Hz$^{-1}$ (Oke 1974).

$= -2.5 \log (DN/s)+15.6$; 1 DN $= 31.3$ e$^-$ (Ricker et al. 1981).

[**] Near infrared MASCOT CCD objects (this paper).

[†] IR positions from Becklin and Neugebauer (1975) measured at 2.2μm; uncertainties are estimated to be ±2" in both coordinates (Becklin, Matthews, Neugebauer, and Willner 1978a).

[††] VLA radio position from Brown, Johnston, and Lo (1981).

[#] Optical reference star positions from G. Rieke (private communication).

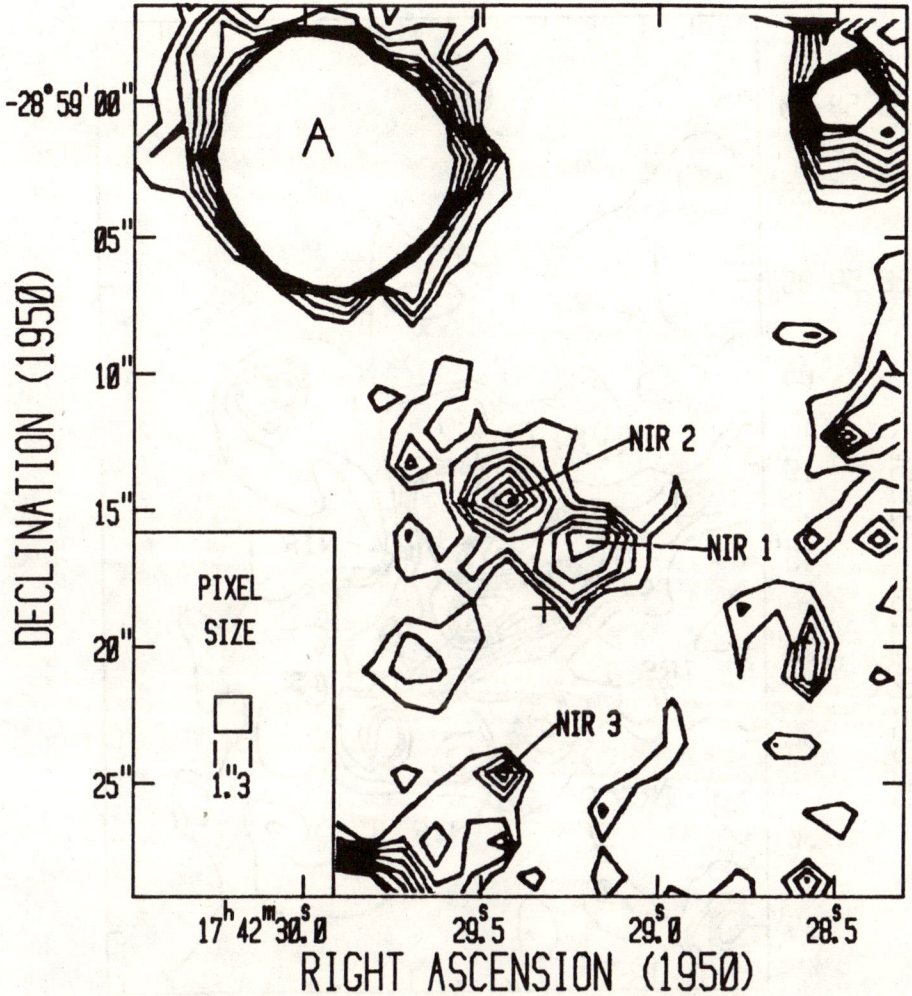

*Figure 1*: *Isophotal contours for a CCD image of the Galactic Center in the passband 9150Å±120Å. The region shown is 30"x30" (~1 pc by 1 pc at 10 kpc), centered near the position of the nonthermal source in Sgr A West[4] (indicated by a cross). It is taken from the central portion of an image 320"x425" obtained in a 1 hr integration on 5/6 September 1981 using the MASCOT CCD instrument[10] on the CTIO 1.5m telescope. Isophotal contours are shown at intervals of 10 DN (1 DN=31.3 detected electrons). The near IR objects which we discovered (NIR 1, NIR 2, and NIR 3) are not visible on the SRC I survey plate (courtesy of the UK Schmidt Telescope Unit). The bright object with an unfilled center designated "A" is a previously known optical field star which we used for coordinate determinations. The CCD pixel size corresponds to 1".3x1".3 (indicated in the inset).*

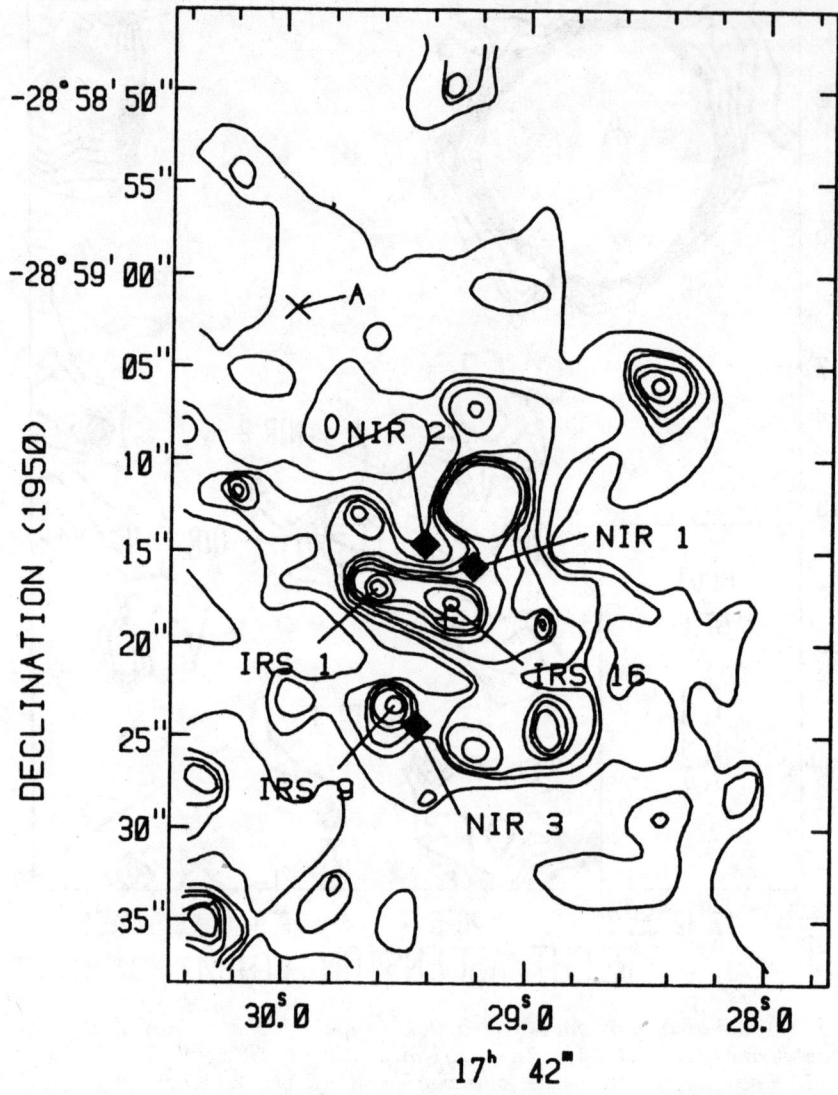

Figure 2: A superposition of the 3 CCD sources (diamond-shaped symbols) detected with the MASCOT near 0.9μm on the 2.2μm contour map of Becklin and Neugebauer.[2] The position of the compact nonthermal radio source in Sgr A West is shown by a cross.[4] "A" designates the optical star used to establish the astrometric grid for both the 2.2μm map and our 0.9μm image.

three 0.9μm sources in our CCD images and the three 2.2μm objects does in fact exist.

It is also interesting to consider the 2.2μm sources which we did not detect. In particular, there are 4 objects comparable to or brighter than IRS 1, 9, and 16 at 2.2μm which are within the region of our CCD images which we have already analyzed. These four objects are IRS 7, 11, 12, and 19.[19] A number of groups[20-22] have argued that these objects are late-type supergiant stars, primarily based on indications for CO absorption at 2.3μm.

Utilizing the spectral models for such stars formulated by Lee[23] we have estimated the expected brightness of these four objects at 9150A. These calculated results are tabulated in Table 2, along with our measured upper limits. From this comparison, it appears reasonable that we would not have detected these 4 objects. (It is possible that superposing all of our images might provide a sufficient gain in sensitivity that we could begin to see one or more of these sources; such work is currently in progress.)

Table 2: Estimated Brightness Limits for Additional Galactic Center Infrared Sources at 0.9μm

| IRS Designation[*] | Upper Limits from this work[†] (Magnitudes at 9150A) | Expected Brightness$_o$[††],[**] (Magnitudes at 9150A) |
|---|---|---|
| 7  | >20.7 | $m_{9150} \gtrsim 20.9$ |
| 11 | >20.7 | $21 \gtrsim m_{9150} \gtrsim 22.6$ |
| 12 | >20.7 | $21 \gtrsim m_{9150} \gtrsim 21.7$ |
| 19 | >20.7 | $20.9 \gtrsim m_{9150} \gtrsim 21.9$ |

[*] Source names from Becklin and Neugebauer (1975).

[**] Magnitudes ($m_{9150}$) are defined according to the convention of Oke (1974).

[†] Obtained with the MASCOT on the 1.5m CTIO telescope (this work). Two sigma upper limits are listed.

[††] Calculated from 1.2μm/2.2μm flux measurements by the formalism of Lee (1970), using measurements given by Neugebauer et al. 1976; Soifer, Russell, and Merrill 1976; and Treffers et al. 1976.

## DISCUSSION

In view of the strong arguments that IRS 16 actually coincides with the Galactic Center[2,4,15] the possibility that NIR 1 might be identified with IRS 16 is a very exciting one. Furthermore, it is also interesting to note that Lacy et al.[24] have pointed out that IRS 1 (which is close to NIR 2) is coincident with a Ne II cloud with $v_{LSR} \simeq 0$, so that it, too, must be considered as a candidate for the location of the dynamic center of the Galaxy.

As to the nature of IRS 16, Becklin et al.[19] have argued, based on its extended size (~2") and its lack of CO absorption, that it could be a cluster of ~$10^2$ giant K5-M0 stars. We have calculated that the continuum luminosity of such an assemblage would in fact result in an object with $m_{9150Å} \simeq +19$, as observed for NIR 1. Furthermore, we cannot exclude the possibility that the objects we have detected are extended on a 1"-2" scale.

Another possibility is that the flux observed from the three NIR objects is line emission from compact H II regions. For example, the [S III] $^1D$–$^3P$ (J=2-1) transition at 9069Å,[25,26] which is commonly observed as a bright feature in H II regions could produce the flux we have detected.

## CONCLUSIONS AND FURTHER WORK

Based on our limited knowledge of the three "far red" sources reported here, it is not yet possible to establish their identity uniquely. Spectroscopic studies with the MASCOT should be able to establish whether they are H II regions emitting strongly in [S III] at 9069Å in the passband reported here. If so, we would expect to also see another [S III] line at 9532Å and possibly several hydrogen Paschen lines in the 0.9-1.1μm range. Further measurements of the radio, optical, and infrared astrometric standards in this portion of the sky, with a goal of establishing more accurate (0".1) correspondences, are essential. Observing programs for generating IR maps at 1.6 and 2.2μm with better than 1" resolution have already been initiated,[27] and will be of great value in clarifying source correspondences in different wavelength ranges. In fact, deeper images with CCDs in the 0.8-1.0μm band should be very helpful in resolving positional uncertainties in the 1-2μm maps, as soon as the suspected supergiant stars (e.g., IRS 7, 11, 12, and 19) can be detected in such CCD images. The results of such observations should become available quite soon, so that we may expect to establish a much more sound observational basis for understanding the detailed structure of the nucleus of our Galaxy in the near future.

## ACKNOWLEDGEMENTS

We gratefully acknowledge useful discussions with Drs. Eric Becklin, George Clark, John Doty, Josh Grindlay, Susan Kleinmann, Barry Lasker, George Rieke, and Steve Willner regarding the interpretation of our data. Dan Dewey and Doug Mink played a key role in developing the MASCOT hardware and software. At CTIO, Dr. Pat Osmer and Mr. Oscar Saa were extremely helpful in facilitating our observations, as also were Messrs. Arturo Gomez and Ricardo Vanegas. Mr. J.C. Golson at Kitt Peak kindly loaned us a number of filters for use with the MASCOT. During the MASCOT project, we have consulted with our friends and colleagues at the C.S. Draper Laboratory, at Texas Instruments, and at the Jet Propulsion Laboratory, including P. Greiff, H. Huemmler, M. Blouke, J. Carlo, H. Hosack, D. McGrath, J. Janesick, and F. Vescelus. Dean R.A. Alberty at MIT provided timely financial support and encouragement which made these observations possible. This work was supported in part by the National Aeronautics and Space Administration under grants NGL22-009-015 and NGL22-009-638.

## REFERENCES

1. Rieke, G.H. and Low, F.J., Ap. J., 184, 415 (1973).
2. Becklin, E.E. and Neugebauer, G., Ap. J. (Letters), 200, L71 (1975).
3. Lacy, J.H., Baas, F., Townes, C.H., and Geballe, T.R., Ap. J. (Letters), 227, L17 (1979).
4. Brown, R.L., Johnston, K.J. and Lo, K.Y., Ap. J., 250, 155 (1981).
5. Watson, M.G., Willingale, R., Grindlay, J.E., and Hertz, P., Ap. J., 250, 142 (1981).
6. Matteson, J., presented at the Workshop on the Galactic Center (Pasadena, January 1982).
7. Spinrad, H., Liebert, J., Smith, H.E., Schweizer, F. and Kuhi, L.V., Astrophys. J., 194, 265 (1971).
8. Becklin, E.E., Matthews, K., Neugebauer, G., and Willner, S.P., Ap. J., 220, 831 (1978).
9. Liller, W. and Alcaino, G., A.J., 85, 532 (1980).
10. Ricker, G.R., Bautz, M.W., Dewey, D., Meyer, S.S., in Solid State Imagers for Astronomy, eds. J. Geary and D. Latham (SPIE 290, Bellingham, Washington, 1981) p. 190.
11. Hynecek, J. IEEE Trans. on Electron Devices, ED-28, 483 (1981).
12. Janesick, J., Hynecek, J., and Blouke, M., in Solid State Imagers for Astronomy eds. J. Geary and D. Latham (SPIE, Bellingham, Washington, 1981).
13. Meyer, S.S. and Ricker, G.R. in Applications of Digital Image Processing to Astronomy, ed. Elliott, D. (NASA/JPL Proceedings S.P.I.E. 264, Bellingham, Washington, 1980) p. 38.

14. Dewey, D. and Ricker, G.R. in *Applications of Digital Image Processing to Astronomy* ed. Elliott, D. (NASA/JPL Proceedings S.P.I.E. 264 Bellingham, Washington, 1980) p. 42.
15. Oort, J.H., *Ann. Rev. Astron. Astrophys.*, 15, 295 (1977).
16. Becklin, E.E. and Neugebauer, G., *Ap. J.*, 151, 145 (1968).
17. Oke, J.B. 1974, *Ap. J. Suppl.*, 27, 21 (1974).
18. Strom, K.M. "Standard Stars for Intensified Image Dissector Scanner Observations" (KPNO Memorandum, unpublished) (1977).
19. Becklin, E.E., Matthews, K., Neugebauer, G., and Willner, S.P., *Ap. J.*, 219, 121 (1978).
20. Treffers, R.R., Fink, K.U., Larson, H.P., and Gautier, T.N. III, *Ap. J. (Letters)*, 209, L115 (1976).
21. Soifer, B.T., Russell, R.W., and Merrill, K.M., *Ap. J. (Letters)*, 207, L83 (1976).
22. Neugebauer, G., Becklin, E.E., Beckwith, S., Matthews, K. and Wynn-Williams, C.G., *Ap. J. (Letters)*, 205, L139 (1976).
23. Lee, T.A., *Ap. J.*, 162, 217 (1970).
24. Lacy, J.H., Townes, C.H., Geballe, T.R. and Hollenbach, D.J., *Ap. J.*, 241, 132 (1980).
25. Allen, L.H. and Liller, W., *Ap. J.*, 130, 45 (1959).
26. Johnson, H.L. in *Nebulae and Interstellar Matter*, eds. B.M. Middlehurst and L.H. Aller (University of Chicago Press: Chicago, 1968).
27. Storey, J.W.V., presented at the *Workshop on the Galactic Center* (Pasadena, January 1982).

# THE POSITION OF THE INFRARED SOURCE IRS 16 IN THE GALACTIC CENTER REGION RELATIVE TO A VISUAL FIELD STAR

G. Neugebauer, K. Matthews, and B. T. Soifer
Palomar Observatory, California Institute of Technology, Pasadena, CA 91125

The position of IRS 16 was measured with respect to the visual star whose position is approximately

$$\alpha(1950) = 17^h 42^m 30^s.0, \quad \delta(1950) = -28°59'01''$$

on 1981 July 25 UT and 1981 Sep 1 UT using the 5m Hale Telescope of Palomar Observatory. Although the source IRS 16 was observed at 2.2 $\mu$m, the visual star could not be isolated at 2 $\mu$m because of confusing sources in its neighborhood.

For this reason, an indirect measuring technique was used. Initially the position in the visual guide field corresponding to the center of the 2.2 $\mu$m diaphragm was determined using a nearly bright star; the 2.2 $\mu$m diaphragm was 1.5'' in diameter. The telescope was then pointed to the Galactic Center and the 2.2 $\mu$m diaphragm centered on the infrared flux from IRS 16. Simultaneously the position of the visual star near IRS 16 was marked in the visual guide field. The telescope was then moved to make the image of the visual star coincide with the previously marked position of the 2.2 $\mu$m diaphragm. This motion, corresponding to the separation of the visual stars and the 2.2 $\mu$m position of IRS 16 was measured on the telescope differential encoders. The telescope was returned to IRS 16, the motion recorded, and the procedure repeated.

In two nights, 18 movements were recorded; the average displacements of IRS 16 from the visual star were:

$$\Delta\alpha = -7.9'', \quad \Delta\delta = -18.2''$$

The statistical uncertainties in the average displacements based on the deviations in the 18 measurements made in both nights are 1 $\sigma = 0.1''$. The systematic errors, which dominate the uncertainties, are harder to assess, but are probably less than or on the order of 0.5''.

In comparison, Becklin and Neugebauer[1] find

$$\Delta\alpha = -9.1'' \quad \Delta\delta = 16.3''$$

with uncertainties of 1.5''; the observations thus formally agree within their uncertainties.

The displacements between the visual star and the two sources found in the CCD frames by Biretto, Lo and Young[2] are:

source A  $\Delta\alpha = -8.7'' \pm 0.1''$  $\Delta\delta = -16.1'' \pm 0.1''$

source B  $\Delta\alpha = -6.6'' \pm 0.1''$  $\Delta\delta = -13.8'' \pm 0.1''$

The position of IRS 16 as determined in the present observations thus differs from that of both sources, although the differences between IRS 16 and source A is less clear given the uncertain nature of the systematic errors.

It is of interest to know if IRS 16 would be expected to have been detected at 0.9 $\mu$m. Becklin et al[3] find that the dereddened magnitudes for IRS 16 are K(2.2 $\mu$m) = 5.6 ± 0.2 and H(1.65 $\mu$m) = 5.7 ± 0.2 mag. A measurement of the colors of IRS 16 with a 1.5'' diameter diaphragm and a north south chop 4'' in length indicates the dereddened magnitude of IRS 16 at 1.25 $\mu$m in a 3.8'' diaphragm would be J = 6.4 ± 0.3 mag. It is thus reasonable that the intrinsic magnitude of IRS 16 at 0.9 $\mu$m would be between 6 and 8. The extinction at 0.9 $\mu$m corresponding to $A_v$ = 30 mag is approximately 13 mag. If the 0.9 $\mu$m magnitude of IRS 16 is 6 mag, it would easily have been seen in the CCD frames while if it were as faint as 8 mag it would, as an extended source, be only marginally detectable. The identification of the sources seen in the CCD frames with IRS16 is thus still unresolved.

We thank J. Carrasco, S. E. Persson, J. Lacy, and K. Y. Lo for assistance in this work and many discussions. This work was supported by grants from NSF and NASA.

## REFERENCES

1. Becklin, E. E. and Neugebauer, G., Ap. J. (Letters), 200, L7 (1975).
2. Biretta, J., Lo, K. Y., and Young, P. J., (This volume).
3. Becklin, E. E., Matthews, K., Neugebauer, G. and Willner, S. P., Ap. J., 219, 118 (1978).

Chapter V   Gamma-Ray Observations

OBSERVATIONS OF CONTINUUM X-RAY AND
GAMMA-RAY EMISSION FROM THE GALACTIC CENTER

James L. Matteson
Center for Astrophysics and Space Sciences, C-011
University of California, San Diego
La Jolla, CA 92093 USA

ABSTRACT

Observations of the X-ray and gamma-ray continuum emission from the galactic center region are reviewed with emphasis on those which are most relevant to the galactic nucleus. X-ray emission in the 0.5 to 4.5 keV band has been discovered from Sgr A West, the site of the galactic nucleus, in observations with 1' resolution. The luminosity is $\sim 1.5 \times 10^{35}$ erg/sec. Observations at higher energy have not resolved the galactic nucleus from nearby sources, but have discovered a hard, variable source with a position uncertainty of a few degrees which includes the galactic nucleus. Its maximum luminosity in the 10 keV to 10 MeV range, $\sim 3 \times 10^{38}$ erg/sec, is the largest of any galactic source in this energy range. The unique nature of this source makes it likely that it is the galactic nucleus itself. Its variability implies a size of < 0.2 pc, which is consistent with the compact clouds observed near the galactic nucleus in the infrared lines of Ne II. A power law photon spectrum which connects the $\sim 1$ keV to $\sim 2$ MeV data must have an average slope of between -0.5 and -1.0. At $\sim 2$ MeV the spectrum must break to an average slope of < -2.5 which continues up to several hundred MeV.

INTRODUCTION

X-ray and gamma-ray continuum emission is expected from objects in which thermal processes with temperatures > $10^7$ K or non-thermal processes by relativistic electrons occur. In the 1 to 10 keV range observations of the galactic center region with 1' angular resolution have resolved point sources. From 10 keV to 10 MeV the angular resolution achieved to date varies from $\sim 1°$ to $\sim 40°$, so individual sources have not necessarily been resolved. However, the large variability of the galactic center flux between 50 keV and 1 MeV leads to the conclusion that most of the unresolved emission when the intensity is large is associated with a single object that is unique in the Galaxy. Observations at > 100 MeV have achieved an angular resolution of a few degrees and isolated a discrete source within one degree of the galactic nucleus.

In this paper the observations of the galactic center region which are relevant to the galactic nucleus itself are reviewed. Emphasis is placed on the observations with the best angular and temporal resolution and sensitivity. The galactic nucleus is taken to be the $\sim 10"$ region which includes a) the point non-thermal radio source in Sgr A West [1], b) the infrared source IRS 16 [2] and c) the dynamical center of the group of compact clouds seen in Ne II infrared fine structure emission [3]. Luminosities are calculated assuming a 10 kpc distance to the source(s) of the flux.

## OBSERVATIONS

a)   1 to 10 keV

X-ray emission from the galactic center was discovered by the Uhuru satellite in 1971 [4]. A source region $1°$ in extent, called GCX, was indicated by scans with a $0.5°$ collimator. It contained the galactic nucleus and had a 1 to 10 keV luminosity of $1.1 \times 10^{37}$ erg/sec. By the late 1970's instruments using rotating modulation collimators and coded masks had produced images of the galactic center. These results had angular resolution from 2.5' to 15'. Skinner [5] has summarized the results. Six sources were found within $1.5°$ of the galactic nucleus. Their luminosity ranged from $2 \times 10^{36}$ to $2 \times 10^{37}$ erg/sec except that one had a peak luminosity of $4 \times 10^{38}$ erg/sec. The latter, A1742-289, is a transient source which increased dramatically in February 1975 and then faded to $\sim 2 \times 10^{36}$ erg/sec by June 1976. This source is located $\sim 1'$ from the galactic nucleus, but is thought to be a typical X-ray transient and not associated with the galactic nucleus [5,6]. X-rays from the galactic nucleus were not detected and a limit of $2 \times 10^{36}$ erg/sec was placed on its luminosity [5].

The second High Energy Astronomical Observatory, also known as the HEAO-2 and the Einstein Observatory, was launched into earth orbit in 1978. It carried a focusing X-ray telescope which operated over the $\sim 0.5$ to 4.5 keV range and had a sensitivity to point sources that was a factor of 100 to 1000 times better than previous instruments. Watson et al. [7] used it in 1979 to image the galactic center region with a $\sim 1'$ resolution. They found 12 sources and unresolved diffuse emission within $\sim 0.5°$ of the galactic nucleus. In particular the brightest of these sources, their source (3), was coincident with the galactic nucleus. Its luminosity was $1.5 \times 10^{35}$ erg/sec and was constant to $\sim 2$ percent in two observations separated by six months. The luminosity must be taken with caution since the spectrum and column density have not yet been determined and therefore the conversion from measured counts to source luminosity may be in error by as much as a factor of 10. $1.5 \times 10^{35}$ erg/sec is only $\sim 10^{-3}$ of the 1 to 10 keV luminosity of the most luminous galactic X-ray sources.

b)   10 to 100 keV

There are several factors which have made observational progress on the galactic center difficult at energies above $\sim 10$ keV. The source fluxes and the detector collecting areas are typically lower than at lower energies. Since a practical focusing collector is not available above $\sim 10$ keV, most of the advantages of the Einstein Observatory are not yet possible. The 10 to 100 keV observations of the galactic center to date have nearly all been conducted with mechanical collimators based on photoelectric absorption which are similar in concept to those used by the Uhuru.

Balloon observations in the early 1970's with 1.5° angular resolution [8] resulted in the discovery of several sources of 17 to 50 keV X-rays within 10° of the galactic center. X-ray emission from the galactic nucleus or the GCX region was not detected. The OSO-8 satellite made a 5 day observation of the galactic center in September 1978. The high-energy X-ray detector, which had 5° angular resolution, detected 20 to 100 keV X-rays from a source within $\sim 2°$ of GCX [9]. The source's position uncertainty contour included GCX. A photon power law with a slope of $\sim -2.3$ fit the data and required a 10 to 100 keV luminosity of $\sim 3 \times 10^{37}$ erg/sec.

The UCSD/MIT experiment on the HEAO-1 satellite, launched in August 1977, provided the first opportunity to simultaneously achieve good angular resolution, $\sim 1.6°$ in the 13 to 200 keV range, good sensitivity, $10^{36}$ erg/sec for a source at 10 kpc, and a complete survey of the sky. The HEAO-1 observed the galactic center once each six months. Each observation consisted of a succession of scanning exposures, one exposure every $\sim 30$ minutes for $\sim 30$ days as the scan plane motion carried the apertures over the galactic center region. 13 to 100 keV data from the Observation 1, in September 1977, are shown in Figure 1. Although there were occasionally several X-ray sources in the aperture at once in this crowded region of the sky, the stronger ones were easily separated with data fitting procedures and their flux was determined [11]. The a priori knowledge of source positions from the 1 to 10 keV X-ray surveys greatly simplified this process. The location and size of GCX in the HEAO-1 data is currently being studied. The work to date indicates the size is < 0.5° and the location is within 0.5° of the galactic nucleus. Referring to Figure 1, it is clear that in the 13 to 50 keV band GCX produced a flux that was typical of the $\sim 10$ strongest sources within 10° of the galactic center. It accounted for $\sim 14$ percent of the total flux of these sources in September 1977. However, in the 50 to 100 keV band GCX was the strongest source at this time and produced $\sim 23$ percent of the total flux. Above $\sim 50$ keV GCX had the hardest spectrum of the galactic center sources and the trend indicated by these data would result in GCX accounting for 50 percent of the flux within 10° of the galactic center at 300 keV.

This effect is clear in Figure 2 which shows the 13 to 180 keV spectra of GCX and the sum of the resolved sources within 10° of the galactic nucleus which were measured by the HEAO-1. Power law photon spectrum fits to GCX done by eye give a slope of -2.1, -2.6 and -2.1 for Observations 1, 2 and 3, respectively. However, in Observation 3 the data require a flatter spectrum above 50 keV with a slope of -1.4. At this time GCX accounted for $\sim 42$ percent of the total galactic center flux at 100 keV. These results also show that GCX is variable by a factor of three at 100 keV on a 6-month time scale. However, the variation is only $\sim 15$ percent at 20 keV. The 10 to 100 keV luminosity of GCX was $6 \times 10^{37}$, $4 \times 10^{37}$ and $7 \times 10^{37}$ erg/sec in the three observations, respectively.

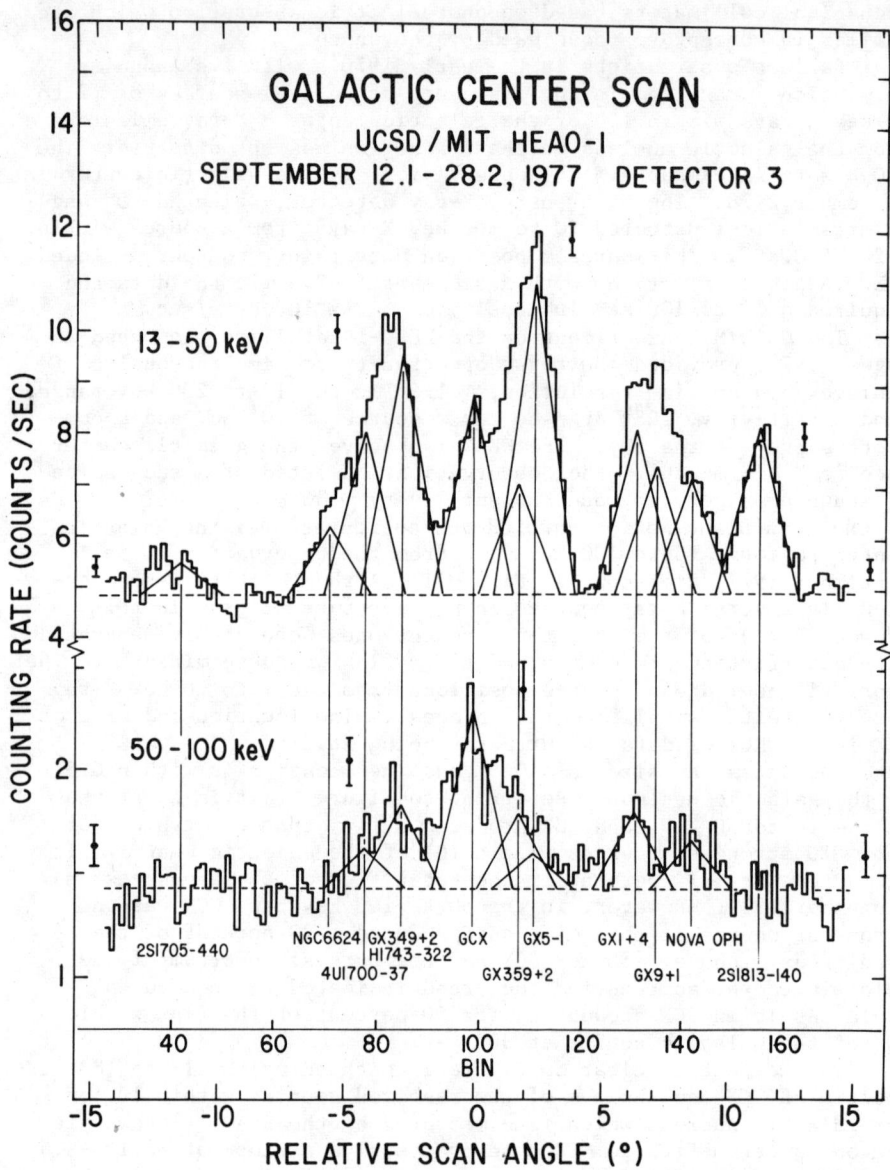

Figure 1. HEAO-1 scans of the galactic center region in September 1977. The data are fitted by a background, indicated by a dashed line, and the flux from discrete sources, indicated by triangles. The triangular shape is due to the transmission of the 1.6° FWHM collimator. In the 13 to 50 keV range the brightness of GCX is typical of the other sources in the region. However, in the 50 to 100 keV range GCX is the brightest source in September 1977.

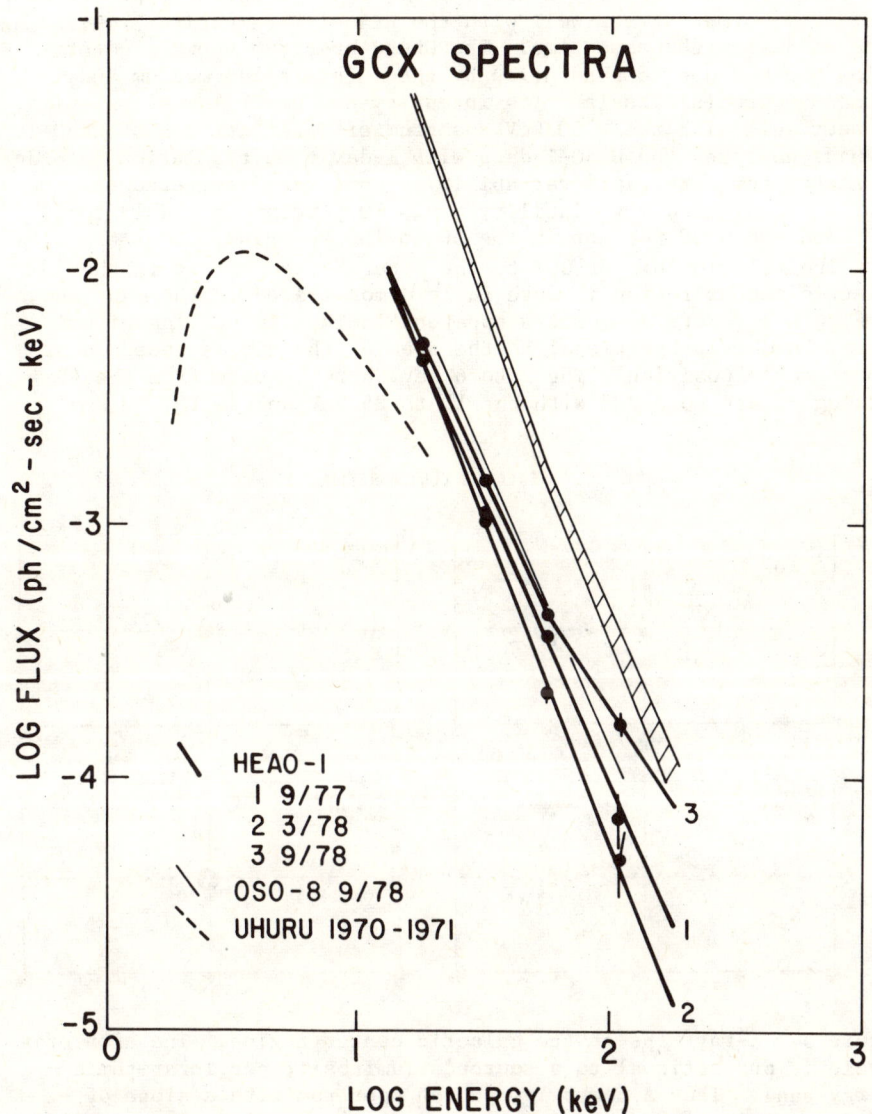

Figure 2. X-ray spectra of GCX. The 13 to 180 keV spectra measured by the HEAO-1 on three occasions are shown as the bold lines. Power law fits have been drawn by eye. A factor of three variability in 6 months at 100 keV is indicated. The UHURU spectrum is cut off below ∼ 3 keV due to absorption in the interstellar medium and perhaps the source itself. The cross-hatched area represents the range of variation of the sum of the sources within 10° of the galactic center that were resolved by the HEAO-1 on the three occasions.

Also shown in Figure 2 are the OSO-8 [9] and Uhuru [4] spectra of GCX. The former agrees well with the HEAO-1 Observation 3 which was done at nearly the same time. The Uhuru spectrum shows a cutoff below ∿ 3 keV due to absorption by the interstellar medium and perhaps source(s) itself. Its intensity and power law slope, -1.4, connect well with the > 50 keV spectrum of Observation 3 of HEAO-1. We have analyzed the HEAO-1 data with 1-day time resolution in order to search for more rapid variability. None was discovered and the limits on day-to-day variability are ∿ 10 percent in the 13 to 50 keV band and ∿ 20 percent in the 50 to 100 keV band.

The relationship of GCX to the other X-ray sources in the galactic center region is more fully demonstrated in the maps shown in Figure 3. Here a source's apparent luminosity per logarithmic energy band is proportional to the area of the circle centered at the source's position. The 2 to 6 keV Uhuru results from the 4U Catalog [10] are compared with the 13 to 25 keV and 80 to 180 keV

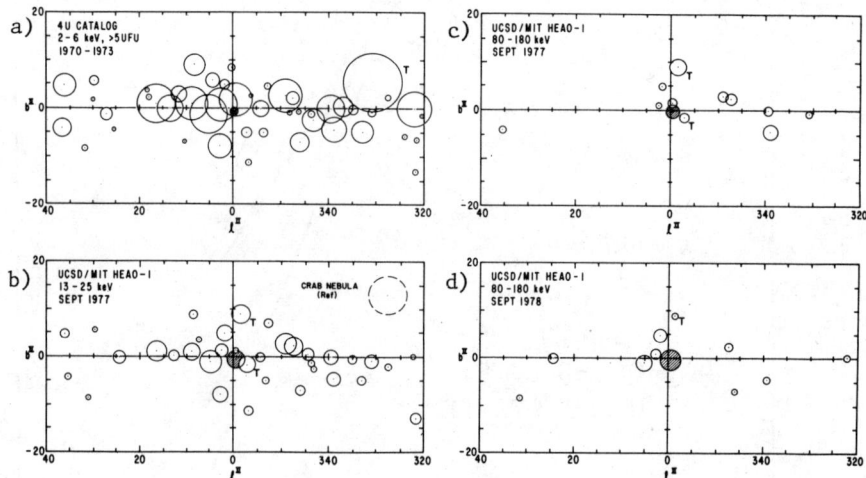

Figure 3. X-ray maps of the galactic center region. The area of a circle is proportional to a source's luminosity per logarithmic energy band. Thus a power law photon spectrum with a slope of -2 would result in the same size circle at all energies. GCX is indicated by the cross-hatched region. "T" indicates a transient source. a) 2 to 6 keV range in 1970 to 1973. The galactic center region is dominated by ∿ 20 bright sources and GCX is relatively faint. b) 13 to 25 keV range in September 1977. Most of the sources are relatively less luminous than in the 2 to 6 keV range, but the brightness of GCX is typical of the other sources. c) 80 to 180 keV range in September 1977. GCX is the brightest source except for the transient, Nova Oph 1977. d) 80 to 180 keV range in September 1978. The nova and the other transient have faded. GCX has further brightened and is the brightest source.

results of the HEAO-1 [11]. A source with a power law photon spectrum of slope -2 would appear with the same size circle in all maps. Referring to Figure 4a, GCX, indicated by the cross-hatched area, is dwarfed by the bright 2-6 keV sources. Most of the sources are fainter in the 13 to 25 keV range in September 1977, but GCX is brighter, c.f. Figure 4b. In the 80 to 180 keV band in September 1977, c.f. Figure 4c, GCX is the brightest source with the exception of Nova Oph 1977, an optical nova which was also a strong X-ray source. In September 1978, c.f. Figure 4d, the nova had faded while GCX brightened to clearly be the brightest 80 to 180 keV source in the galactic center region.

c) 100 keV to 10 MeV

The observations of the galactic center region in the 100 keV to 10 MeV range have also been made with instruments that use mechanical collimators. Since Compton scattering is the dominant photon interaction mechanism at these energies, collimation is difficult and requires large, massive devices. Therefore, practical instrument design has resulted in angular resolution in the $15°$ to $40°$ range. (Compton telescope instruments can obtain an angular resolution of a few degrees in the 1 to 20 MeV range by directly using the Compton scattering process to obtain directional information. However, an observation of a discrete source in the galactic center region has not yet been reported from these instruments).

Measurements of the galactic center region spectrum in the $\sim 70$ to $\sim 500$ keV range are shown in Figure 4. All these results are from instruments with $13°$ to $16°$ angular resolution, except for the original instrument of the Rice group [12] and the HEAO-3 [13], which had $24°$ and $30°$ resolution, respectively. Each measurement shown in Figure 4 typically includes contributions from 5 to 10 80 to 180 keV sources resolved by the HEAO-1 in the galactic center region. For simplicity, only the power law fits to the data are shown. The 1974 Rice observation [15], number 2 in Figure 4, was centered on the X-ray source GX 1+4 which is located $5°$ from the galactic nucleus. Since the instrument's relative response to a source $5°$ off axis is only $\sim 0.62$, the spectrum in Figure 4 must be increased by a factor of 1.63 (= 1/0.62) if it is interpreted as due to a source located at the galactic nucleus. With this adjustment the measurements shown in Figure 4 cluster between $1.8 \times 10^{-4}$ and $4 \times 10^{-4}$ ph/cm$^2$-sec-keV at 100 keV. However, they show a larger amount of variability at higher energies. This is particularly convincing in the cases where the same instrument made multiple observations. At 300 keV the HEAO-1 observed a factor of 4 decrease in six months and the HEAO-3 observed a factor of 8 decrease in six months [14]. There is a trend for the spectrum to soften as the intensity decreases. The HEAO-3 measurements represent the extreme intensity states, with the 100 keV to 500 keV luminosity decreasing from $\sim 7 \times 10^{37}$ to $\sim 2 \times 10^{37}$ erg/sec while the power law slope decreased from -2.1 to -3.2. The HEAO-1 scanning observations show that the source(s) of the continuum are located within $5°$ of the galactic nucleus and have an angular extent

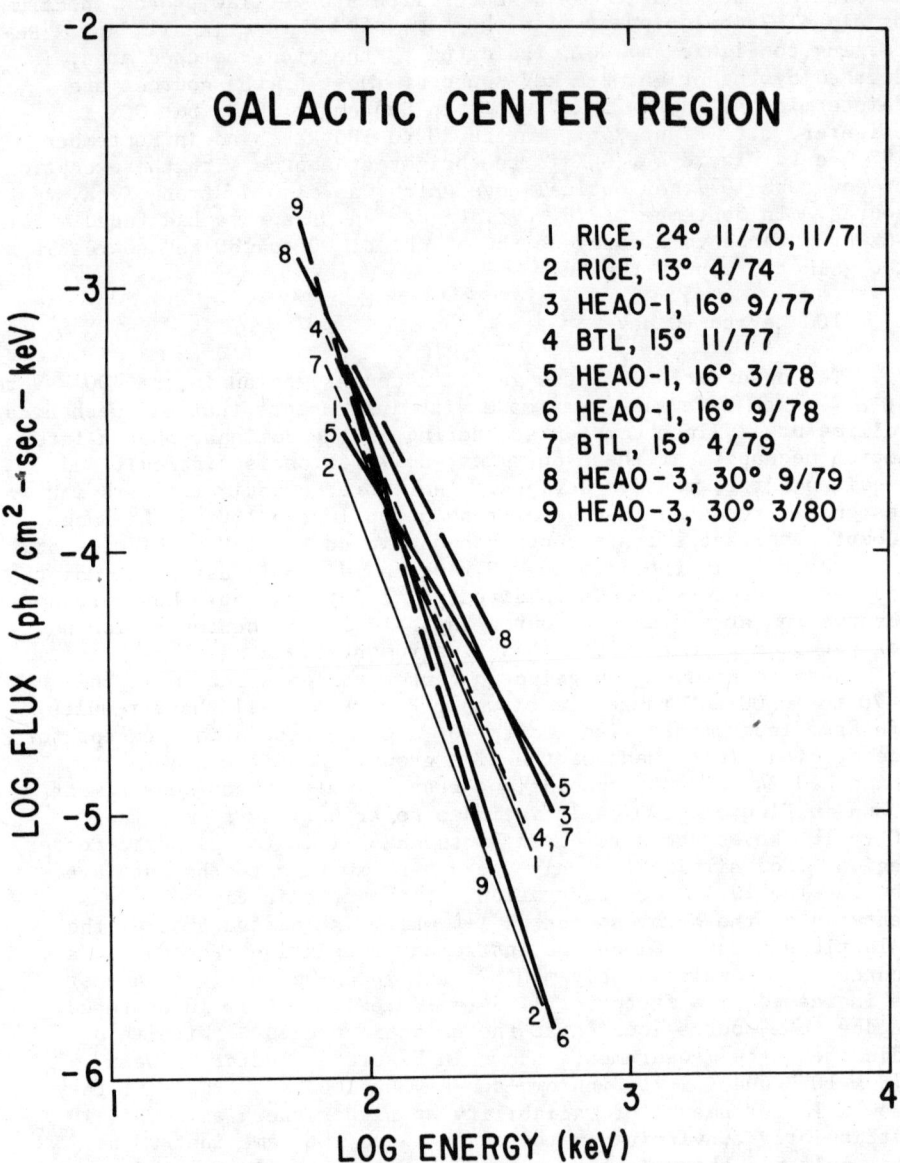

Figure 4. Spectra of the galactic center region. Since the instruments' angular resolution was in the 13° to 30° range, individual sources were not necessarily resolved. The Rice observation in 4/74 should be increased by a factor of 1.63, as explained in the text, in order to give the flux from a source located at the galactic nucleus. Variability of up to a factor of 8 in six months is indicated at 300 keV and there is a clear trend for the spectrum to soften as the intensity decreases. BTL indicates the Bell/Sandia group [23,24].

of < 5°.

Above 500 keV there are very few observations. The HEAO-1 results from September 1977 are shown in Figure 5. The spectrum hardens above 300 keV with the power law slope changing from -2.4 to $\sim$ -1.4. Above $\sim$ 2 MeV the spectrum softens such that the slope is < -2.5. The HEAO-1 data above 1 MeV are from the High Energy Detector (HED) which had a 43° field of view while those below 1 MeV are from the Medium Energy Detector (MED) which had a 16° field of view. Since they agree at $\sim$ 1 MeV, most of the HED flux was from the source(s) detected by the MED. The 500 keV to 10 MeV luminosity implied by the HEAO-1 spectrum is $\sim$ 2 x $10^{38}$ erg/sec. The 1974 Rice spectrum [15] is also shown in Figure 5. After it is adjusted as described above, it is a factor of $\sim$ 3 below that of the HEAO-1, in the 100 keV to 1 MeV range, but above 1 MeV the two spectra are in agreement. The > 500 keV variability implied by these results is $\sim$ 7 x $10^{37}$ erg/sec.

d) 10 MeV to 1 GeV

Although the galactic center region has been observed in the 10 MeV to 100 MeV range by several spark chamber instruments in the past ten years [16,17,18], only a diffuse galactic flux and no point sources have been detected. Above 100 MeV the extensive observations of the COS-B satellite from 1975 to 1978 have resulted in the discovery of 25 discrete sources [19]. One of these, 2CG359-00, has a 1° position error radius which includes the galactic nucleus and has a luminosity of $\sim$ 8 x $10^{36}$ erg/sec.

## DISCUSSION

The observations of the galactic center region are summarized in Table 1 and discussed in more detail below.

The galactic nucleus has not been resolved from the other galactic center region X-ray and gamma-ray sources above 4.5 keV. Therefore, the observed fluxes in this energy range can only be strictly interpreted as upper limits to the galactic nucleus. However, the large variability in the observed flux in the 100 keV to 1 MeV range indicates that in the high flux state a single source is probably the dominant emitter. This is supported by the fact that the difference in the galactic center region high and low flux spectra measured at 120 keV by the HEAO-1 and HEAO-3, $\sim$ 1.2 x $10^{-5}$ ph/cm$^2$-sec-keV, c.f. Figure 4, is nearly equal to the difference in the GCX high and low flux spectra measured by the HEAO-1 at 120 keV, $\sim$ 1.0 x $10^{-5}$ ph/cm$^2$-sec-keV, c.f. Figure 2. The peak luminosity in the 100 keV to 1 MeV range, $\sim$ 3 x $10^{38}$ erg/sec or $\sim$ 30 times that of the Crab Nebula, is the largest of any galactic source. Since this unique object is located within a few degrees of the galactic nucleus, it is likely that it is the galactic nucleus itself.

The situation in the 10 to 100 keV range is more difficult, c.f. Figure 2. Above 50 keV the large variability of GCX indicates that the arguments given above are still valid. However, at lower energies the variability is small, $\sim$ 20 percent, and the fluxes of

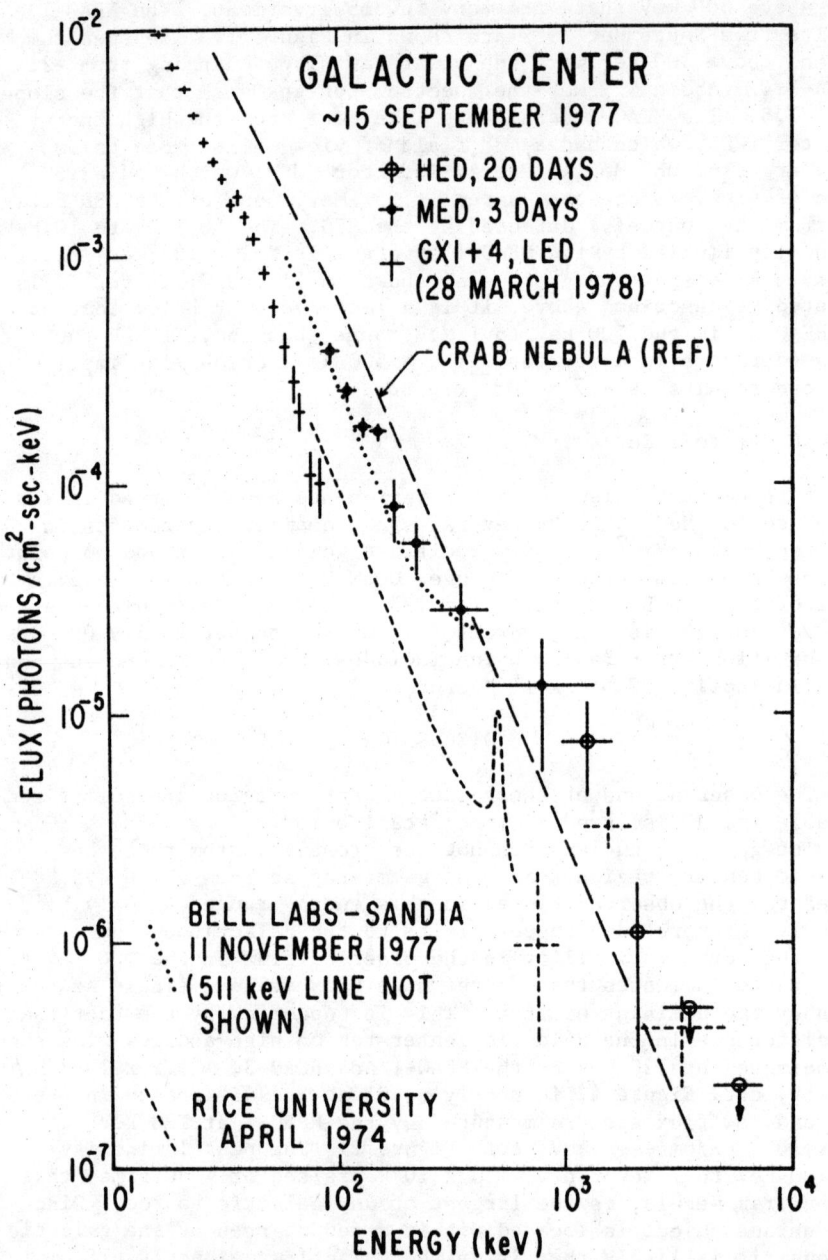

Figure 5. The spectrum of the galactic center region as measured by the HEAO-1 in September 1977. These results show that above ~ 300 keV the spectrum hardens and that this continues up to a cutoff at ~ 2 MeV. For comparison, the spectra of the Crab Nebula, the galactic center region on two other occasions and GX 1+4, a bright X-ray pulsator located 5° from the galactic nucleus, are shown.

Table I  Summary of Observations of the Galactic Center Region

| Log Energy (keV) | 0-1 | 1-2 | 2-4 * | 5-6 |
|---|---|---|---|---|
| Angular Resolution | ~1' | ~1° | ~15° | ~1° |
| Position of Source(s) | Sgr A West | within 0.5° of Sgr A West | within 5° of Sgr A West | within 1° of Sgr A West |
| Angular Size of Sources | <1' | <0.5° (<3")** | <5° (<3")** | <1° |
| Variability | < few percent in 6 months | factor of 3 decrease at 100 keV in 6 months | factor of 8 decrease at 300 keV in 6 months | no data |
| Maximum Luminosity (erg/sec) | $2 \times 10^{35}$ | $7 \times 10^{37}$ | $3 \times 10^{38}$ | $8 \times 10^{36}$ |
| Variable Luminosity (erg/sec) | $<4 \times 10^{33}$ | $3 \times 10^{37}$ | $1.4 \times 10^{38}$ | --- |

\* No discrete sources have been detected in the galactic center region in the 10 to 100 MeV range.

\*\* Based on observed variability.

the ~ 10 sources in the GCX region are unknown. Thus, we have no indication of the flux from the galactic nucleus except that it cannot exceed that of GCX. The > 100 MeV flux of 2CG359-00 is typical of the ~ 24 galactic sources discovered by the COS-B and there is no information available on its variability. Therefore 2CG359-00 does not appear to be a unique galactic object and it cannot be confidently associated with the galactic nucleus at this time.

With these considerations the picture of the X-ray and gamma-ray emission from the galactic nucleus is as follows. The luminosity per energy decade increases from ~ $2 \times 10^{35}$ erg/sec in the 1 to 10 keV band to ~ $2 \times 10^{38}$ erg/sec in the 100 keV to 1 MeV band. For a power law photon spectrum this requires an average slope of ~ -0.5 over the ~ 1 keV to ~ 1 MeV energy range. If the 1 to 10 keV luminosity is 10 times greater, as may result from uncertainties in the low energy X-ray absorption in the spectrum of Sgr A West [7], then the average power law slope is $\sim -1.0$. This spectrum continues to $\sim 2$ MeV. In the $\sim 2$ to 10 MeV range the spectrum steepens to a slope of $< -2.5$. From 2 MeV to $> 100$ MeV the average slope must be $< -2.5$ in order to not exceed the $> 100$ MeV flux of 2CG359-00.

Below $\sim 2$ MeV this spectrum is the hardest of any known X-ray and gamma-ray source. The steepening in the September 1977 HEAO-1 spectrum above $\sim 2$ MeV, c.f. Figure 5, may be interpreted to imply an effective temperature of $\sim 10^{10}$ K. Since the HEAO-1 data were taken at a time of near maximum luminosity, $\sim 10^{10}$ K is probably the maximum effective temperature which occurs.

This discussion of the spectrum is clearly an oversimplification since it is inconsistent with the HEAO-1 observations in September 1978, c.f. Figures 2 and 4. At that time the $\sim 100$ keV flux from GCX was at a maximum and had a power law slope of -1.4. But above $\sim 150$ keV the total spectrum of the galactic center was below the extrapolated GCX spectrum, requiring a steepening in the GCX spectrum to a slope of $< -3$ at $\sim 150$ keV. The energy of the steepening implies an effective temperature of $\sim 10^9$ K at that time.

The variability ranges from less than a few percent in 6 months in the 1 to 10 keV range to a factor of 3 to 8 in 6 months in the 100 keV to 1 MeV range. The extensive observations from 1977 to 1980, c.f. Figure 4, appear to have resolved intensity maxima near September 1977 and September 1979, indicating an $\sim 2$ year time scale for the overall cycle of variation. The variability on a 6 month time scale implies a source size of $< 0.2$ pc, which is in the range of sizes of the compact clouds near the galactic nucleus[3].

The spectrum and variability results indicate that the highest effective temperature, largest luminosity and shortest lifetime are correlated and thus probably are due to processes which occur in one region. This can be satisfied by source models in which an impulsive, central energy release leads to an initial high temperature, $\sim 10^{10}$ K, high luminosity, $\sim 10^{38}$ erg/sec, and short lifetime, $\sim 6$ months in the central region. The outer regions would be heated to only $\sim 10^7$ K, have a lower luminosity, $\sim 2 \times 10^{35}$ erg/sec, and a lifetime of tens of years or more. A quasi-steady luminosity in the 1 to 10 keV range would result from the integrated effect of many impulsive events. At higher energies the luminosity variation would follow the impulsive events accurately and produce the observed variability.

The discussion presented above is based on the assumption that most of the variable flux from GCX in the $\sim 50$ to $\sim 100$ keV range and from the galactic center region in the $\sim 100$ keV to $\sim 1$ MeV range is from the galactic nucleus itself. This needs to be tested by future observations in the 10 keV to 1 MeV range which can resolve the candidate objects which may be the source(s) of the flux. An angular resolution of $\sim 0.1°$ is required. This can be achieved by balloon-carried instruments that use coded masks or modulation collimators.

The observations of the 511 keV $e^+$-$e^-$ annihilation gamma-ray line from the galactic center region have been summarized by Jacobson[20] and Leventhal[21]. The flux in the line is variable form $\sim 6 \times 10^{-4}$ to $\sim 2 \times 10^{-3}$ ph/cm$^2$-sec. The HEAO-3 observed a factor of 3 decrease in the flux [22] in 6 months and also observed a factor of 8 decrease in the 300 keV continuum emission from the galactic center region over the same time interval [14]. Thus it is tempting to propose that the line and continuum fluxes are correlated. However, the 511 keV line has also been observed at a flux of $\sim 10^{-3}$ ph/cm$^2$-sec when the continuum spectrum was in an intermediate state [23,24], c.f. Figure 4, observations 4 and 7, and when the continuum was in a low state [15], c.f. Figure 4, observation 2. When additional factors such as instrument field of view and the uncertain flux of other sources in the galactic center region are considered it is clear that the determination of the correlation of the line and continuum flux will require a careful systematic program of well controlled observations. This will require long term monitoring with observations each $\sim$ one month for a period of several years. The Gamma-Ray Observatory (GRO), scheduled for launch in 1988, will carry the Oriented Scintillation Spectrometer Experiment which will have 5° angular resolution and be capable of measuring the continuum spectrum with its NaI(Tl) detectors, but will not have the good energy resolution required for precise studies of the 511 keV line. This requires cooled Ge detectors which could be carried on balloons in order to perform the 511 keV line observations necessary to complement the GRO continuum observations.

## CONCLUSION

The galactic nucleus contains an X-ray source with a nearly constant luminosity of $\sim 1.5 \times 10^{35}$ erg/sec in the 0.5 to 4.5 keV band. Observations at higher energies have not yet resolved the galactic nucleus from other candidate objects in the galactic center region. However, observations in the 10 keV to 10 MeV band have discovered a bright, hard and variable source within a few degrees of the galactic nucleus. At times of maximum luminosity, $\sim 4 \times 10^{38}$ erg/sec, the photon power law spectrum slope of the variable component must be in the range from -0.5 to -1.0. The spectrum breaks at $\sim 2$ MeV, implying an effective temperature of $\sim 10^{10}$ K. The source's unique nature - its 100 keV to 1 MeV luminosity is $\sim 30$ times that of the Crab Nebula - suggests that it may be the galactic nucleus itself. Future observations of the galactic center region with $\sim 0.1°$ angular resolution over the $\sim 10$ keV to 1 MeV range are required to test this suggestion. Long term monitoring of the galactic center region is required to determine the presently unclear relation between the X-ray and gamma-ray continuum and the 511 keV $e^+$-$e^-$ annihilation line from the galactic center.

## ACKNOWLEDGEMENTS

R. Proctor and D. Worrall of UCSD and A. Levine of MIT made major contributions to the HEAO-1 results presented here. This work was supported by contract NAS 8-27974.

## REFERENCES

1. R.L. Brown, K.J. Johnston and K.Y Lo, Ap.J., 250, 155 (1981).
2. E.E. Becklin and G. Neugebauer, Ap.J.(Letters), 200, L71 (1975).
3. J.H. Lacy, C.H. Townes, T.R. Geballe and D.J. Hollenbach, Ap.J., 241, 132 (1980).
4. E. Kellogg, H. Gursky, S. Murray, H. Tananbaum and R. Giacconi, Ap.J. (Letters), 169, L99 (1971).
5. G.K. Skinner, Proc. R. Soc. Lond. A., 366, 345 (1979).
6. P.G. Murdin, D.A. Allen, D.C. Morton, A.J. Whelan and R.M.Thomas, M.N.R.A.S., 192, 709 (1980).
7. M.G. Watson, R. Willingale, J.E. Grindlay and P. Hertz, Ap.J., 250, 142 (1981).
8. G.R. Ricker, M. Gerassimenko, J.E. McClintock, S.G. Ryckman and W.H.G. Lewin, Ap.J., 207, 333 (1976).
9. B.R. Dennis, J.H. Beall, E.P. Cutler, C.J. Crannell, J.F. Dolan, K.J. Frost and L.E. Orwig, Ap.J. (Letters), 236 L49 (1980).
10. W. Forman, C. Jones, L. Comimsky, P. Julien, S. Murray, G. Peters, H. Tamambaum and R. Giacconi, Ap.J. Suppl., 38, 357 (1978).
11. A Levine, F. Lang, P. Byrne, B.A. Cooke, C.A. Dobson, J.P. Doty, J.A. Hoffman, S.K. Howe, F.A. Primini, A. Scheepmaker, W.A. Wheaton, W.H.G. Lewin, J.L. Matteson, F.K. Knight, P. Nolan and L.E. Peterson, BAAS, 11, 429 (1979).
12. W.N. Johnson III and R.C. Haymes, Ap.J., 184, 103 (1973).
13. W.A. Mahoney, J.C. Ling, A.S. Jacobson and R.M. Tapphorn, Nucl. Instr. and Meth., 178, 363 (1980).
14. G.R. Riegler, J.C. Ling, W.A. Mahoney, W.A. Wheaton and A.S. Jacobson, in Proceedings of the 17th International Cosmic Ray Conference, Paris, France, July 13-25, 1981, Paper OG H.1-1 (1981).
15. R.C. Haymes, G.D. Walraven, C.A. Meegan, R.D. Hall, F.T. Djuth and D.A. Shelton, Ap.J. 201, 593 (1975).
16. G.H. Share, R.L. Kinzer and N. Seeman, Ap.J., 187, 511 (1974).
17. R.C. Hartman, D.A. Kniffen, D.J. Thompson, C.E. Fichtel, H.B. Ögelman, T. Tümer and M.E. Özel, Ap.J., 230, 597 (1979).
18. D.A. Kniffen, D.L. Bertsch, D.J. Morris, R.A.R. Palmeira and D.K. Rao, Ap.J., 225, 591 (1978).
19. B.N. Swanenburg, K. Bennett, G.F. Bignami, R. Buccheri, P. Caraveo, W. Hermson, G. Kambach, G.C. Lichti, J.L. Masnou, H.A. Mayer-Hasselwander, J.A. Paul, B. Sacco, L. Scarsi, R.D. Wills, Ap.J. (Letters), 243, L69 (1981).
20. A.S. Jacobson, in these proceedings.
21. M. Leventhal, in these proceedings.
22. G.R. Riegler, J.C. Ling, W.A. Mahoney, W.A. Wheaton, J.B. Willett and A.S. Jacobson, Ap.J. (Letters), 248, L13 (1981).
23. M. Leventhal, C.J. MacCallum and P.D. Stang, Ap.J. (Letters), 225, L11 (1978).
24. M. Leventhal, C.J. MacCallum, A.F. Huters and P.D. Stang, Ap.J., 240, 338 (1980).

# OBSERVATIONS OF GAMMA-RAY LINE EMISSION
# FROM THE GALACTIC CENTER REGION

Allan S. Jacobson
Jet Propulsion Laboratory
California Institute of Technology
Pasadena, CA 91109

## ABSTRACT

A series of observations of the Galactic Center 511 keV electron-positron annihilation line spanning a decade has begun to severely constrain theoretical ideas concerning the nature of the source. The fluxes reported range from $(4.18 \pm 1.56) \pm 10^{-3}$ $\gamma/cm^2$-s to less than $0.7 \times 10^{-3}$ $\gamma/cm^2$-s ($2\sigma$ upper limit). Data from high-resolution germanium spectrometers show a narrow (1.6 keV; + 0.9, - 1.6 keV), unredshifted ($E = 510.9 \pm 0.25$ keV) emission feature which varies in intensity by a factor of up to three, in six months or less. The combined results show that, in addition to any possible spatially extended component, a significant fraction of the emission must come from one or a few sources with size $r \lesssim 10^{18}$ cm, temperature $T \lesssim 10^5$ K, and typical luminosity $\sim 10^{37}$ ergs s$^{-1}$. Simple models indicate the mass M of any compact central object is $M < 10^9$ $M_\odot$. There is a strong need for continued monitoring of the variations of this source.

## INTRODUCTION

Low energy gamma-ray astronomy is still in its rudimentary stages and flux intensity measurements are more often than not close to the instrumental detection thresholds. For this reason experimenters are occasionally fooled by the observations and must exercise caution in their interpretations. This, coupled with the fact that there are relatively few groups working in this field, with few opportunities to make observations, results in a rather meager history of observational verifications of reported lines. One strongly notable exception has been the observational history of the .511 MeV positron-electron annihilation radiation line from the direction of the Galactic Center. Until the most recent measurements, which have revealed evidence that the line flux has significantly diminished, every attempted gamma-ray observation has yielded the positive detection of spectral feature around .51 MeV.

## THE MEASUREMENTS

Table 1 lists the measurements reported, with the exception of those new measurements to be presented in these proceedings. The pioneering work was carried out by the group at Rice University, which made several expeditons to the southern hemisphere under the leadership of Robert Haymes[1]. The first report of a line detection was based upon a 1970 flight from Argentina. Figure 1 shows the

TABLE I

## POSITRON ANNIHILATION LINE OBSERVATIONS

| Eo (keV) | FLUX ($10^{-3}$/cm$^2$-s) | APERTURE (FWHM) | DATE | GROUP |
|---|---|---|---|---|
| 476 ± 24 | 1.8 ± 0.5 | 24° | 1971 NOV 20 | RICE UNIV |
| 530 ± 11 | 0.80 ± 0.23 | 13° | 1974 APRIL 2 | |
| 511 | 4.18 ± 1.56 | 50° | 1977 FEB 14, 17 | CESR |
| 510.7 ± 0.5 | 1.22 ± 0.22 | 15° | 1977 NOV 11 | BELL/SANDIA |
| 511 | 2.35 ± 0.71 | 15° | 1979 APRIL 15 | |
| 510.90 ± 0.25 | 1.85 ± 0.21 | 35° | 1979 OCT | JPL |
| 510.1 | 0.65 ± 0.27 | 35° | 1980 MARCH | |

averaged results from two balloon flights. The measurements were made with an actively shielded NaI(Tl) of 75 cm$^2$ area and an aperture of 24 degrees FWHM. The energy resolution at .511 MeV was 14.7 percent. A complete description of the instrument can be found in Johnson and Haymes[1]. A line flux of (1.8 ± .5) x $10^{-3}$ photons/cm$^2$-s, centered at an energy of .476 ± .024 MeV was reported.

Several possible explanations for the existence of this feature and its peak energy were advanced. Among them was the suggestion that the feature was due to the cumulative gravitationally redshifted .511 MeV lines from neutron stars in the Galactic Center region[2], or that the gamma rays resulted from the deexcitation of the 478 keV level and the 431 keV level of Li$^7$ and Be$^7$, respectively, in the interstellar medium[3]. Another possible interpretation was advanced that is particularly worthy of note because of its relevance to present observational and theoretical objectives in the study of the Galactic Center. Leventhal[4], pointed out that positron-electron annihilation can take place either via free electron interactions or through the formation of positronium. In positronium, annihilation can take place from either a singlet or a triplet state, the former resulting in two characteristic .511 MeV gamma rays, and the latter in a three-photon continuum. Figure 2 shows this continuum as it would appear in conjunction with the .511 MeV line resulting from the annihilation of positrons all of which were in positronium atoms, and measured with instruments of different energy resolutions. Because of the existence of the continuum, the peak energy would appear to be shifted to a lower energy when observed with low resolution instruments. This then could explain why a .511 MeV peak was shifted.

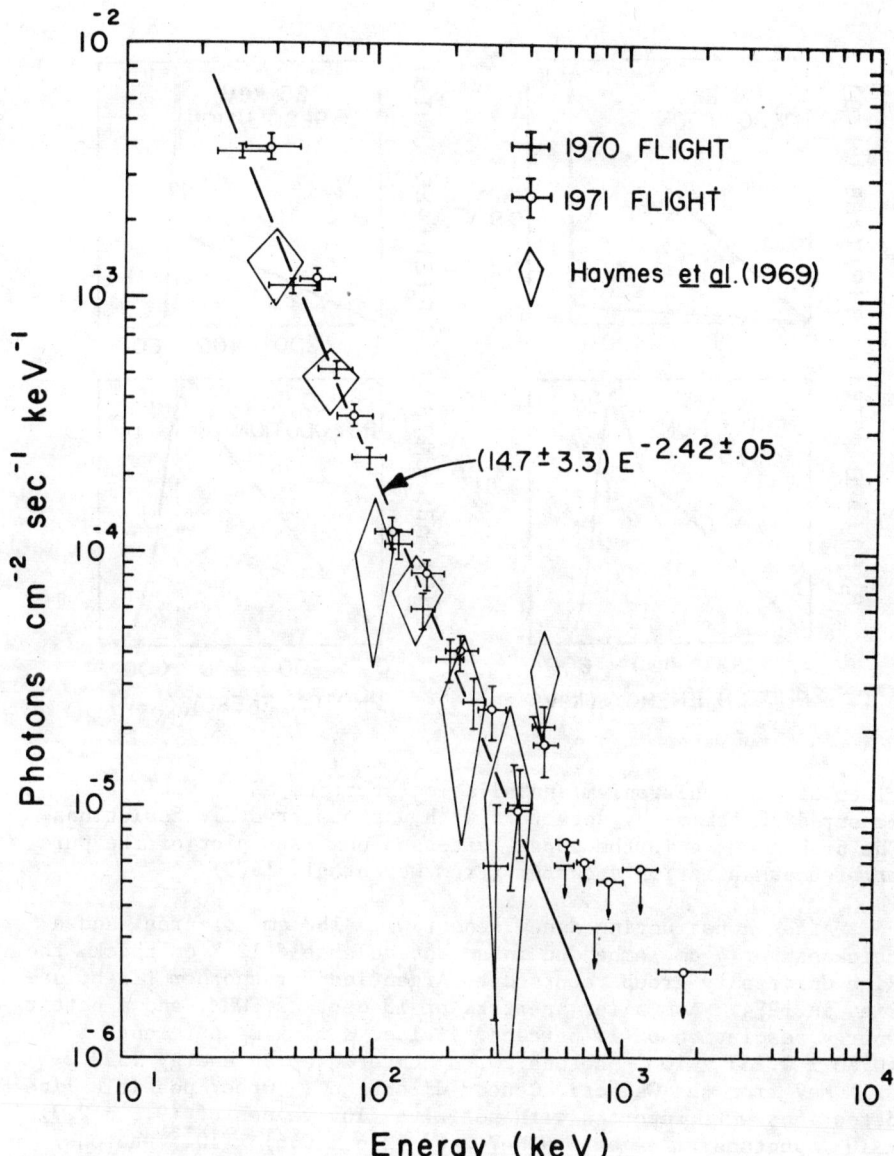

Figure 1
The data from three measurements of the Galactic Center region low energy gamma radiation carried out by the Rice group. A solid state line represents the best fit to the data from the 1970 and 1971 flights. The combined data in the line indicates a measured flux at 476 keV which is 5.3 standard deviations above the power law continuum. (From Johnson and Haymes, 1973).

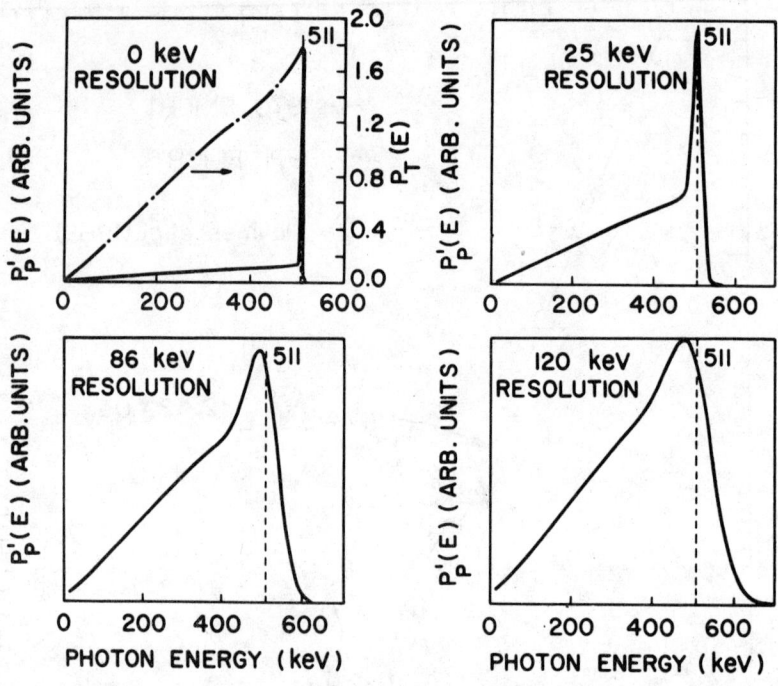

Figure 2
Plots of the positronium annihilation spectrum as it would be measured in gamma-ray detectors with various crystal resolutions. The broken curve in the upper left-hand box is a plot of the pure, triplet annihilation spectrum (From Leventhal, 1973).

After constructing a new sensor with 182 cm² of area, and a thickness of 5 cm, embedded in an active shield 12.7 cm thick, the Rice University group returned to Argentina for another flight series in 1974. A smaller aperture of 13 degrees FWHM, and a better energy resolution of 12 percent, yielded a flux measurement of $(0.80 \pm 0.23) \times 10^{-3}$ photons/cm²-s centered at an energy .530 ± 0.11 MeV from the Galactic Center direction[5]. Other possible line detections were reported with positive flux values of $(9.5 \pm 2.7) \times 10^{-4}$ photons/cm²-s at 4.6 MeV, and $(2.6 \pm 0.6) \times 10^{-3}$ photons/cm²-s at 1.2 - 2.0 MeV.

High spectral resolution germanium detectors came into use for these observations in 1977. On 14 and 17 February 1977, a French/Brazilian consortium flew balloons from Brazil[6] carrying a large Ge(Li) crystal of about 100 cm³ volume in a 5 cm thick NaI(Tℓ) shield, with an aperture of about 25 degrees FWHM. The spectral resolution on the first flight varied between 1 and 2 percent, and was 3.5 percent for the second flight. A weighted average flux value for these two flights was $(4.18 \pm 1.56) \times 10^{-3}$ photons/cm²-s.

The first strongly convincing observation was made in November 1977 by the Bell/Sandia consortium[7], using a 130 cm$^3$ high purity germanium crystal in a 15.2 cm thick NaI(Tℓ) shield. Spectral resolution was .63 percent, and the aperture was 15 degrees FWHM. A long flight was achieved with a total of 17.3 hours of data accumulated, producing a net flux from the Galactic Center direction of $(1.22 \pm .22) \times 10^{-3}$ photons/cm$^2$-s at $510.7 \pm 0.5$ keV. An upper limit of 3.2 keV was set on the line width.

In an attempt to determine what fraction of this radiation might have come from positronium, the 3-photon annihilation spectrum was fit to the continuum measured on the low energy side of the peak. Figure 3 shows the data, and the assumed theoretical spectrum. The chi-square test of this fit indicates that any fraction of positronium from 0 to 100 percent is statistically allowable. However, the application of the F-test indicates that a significantly better fit is achieved if, in addition to a gaussian spectrum for the line and a power law spectrum for the continuum, a third spectral component similar in shape to the three-photon annihilation continuum spectrum is assumed.

Possible additional lines were measured with fluxes of $(7.4 \pm 1.8) \times 10^{-4}$ photons/cm$^2$-s in a feature about 12 keV wide at .170 MeV, $(7.3 \pm 1.5) \times 10^{-4}$ photons/cm$^2$-s in a feature about 10 keV wide at 1.611 MeV, and $(3.1 \pm .8) \times 10^{-3}$ photons/cm$^2$-s in a feature about 500 keV wide at 3.70 MeV.

A second Bell Sandia flight made on 15 April 1979 with the same instrument yielded a flux measurement of $(2.35 \pm .71) \times 10^{-3}$ photons/cm$^2$-s[8]. The low energy continuum had changed slope, and the other possible line features seen in the previous flight were not observed, leading the experimenters to conclude that they were statistical fluctuations. No conclusions could be drawn about a three-photon annihilation spectrum from the data acquired during this flight.

In 1979, the HEAO-3 spacecraft was launched with the JPL gamma-ray spectrometer aboard and provided the only reported obervation of the Galactic Center region made from a satellite. The instrument consisted of 4 large high-purity germanium crystals, each of about 100 cm$^3$ volume in a 6.6 cm thick CsI(Na) anti-coincidence shield. Described fully in Mahoney, et al.[9], it had an aperture of 35 degrees FWHM, and a spectral resolution of .53 percent and 1.11 percent during two observations of the Galactic Center region, in which the spacecraft spin axis was made to coincide with the galactic pole. This took place in the Fall of 1979 and the Spring of 1980. These measurements gave the first information on the spatial extent of the source, and detected a significant time variability in the .511 MeV line emission[10,11].

Figures 4 (a) and (b) show, respectively, the 1979 and 1980 galactic plane scans in the light of .511 MeV emission. A line flux of $(1.85 \pm .21) \times 10^{-3}$ photons/cm$^2$-s was detected in the Fall, emanating from a direction determined to be $\ell^{II} = 3.9 \pm 4.0$ degrees, if a point source is assumed, or $3.5 \pm 4.0$ degrees with an extent of $19 \pm 8$ degrees if an extended source is assumed. The data are consistent with either of these assumptions and it is the

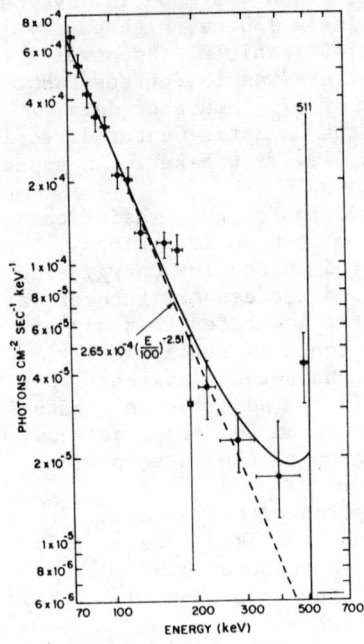

Figure 3

The differential photon spectrum from the Galactic Center region as measured by the Bell/Sandia group. The solid line represents the best fit of a combined power law, gaussian-shaped line and 3-photon annihilation spectra. (From Leventhal, et al., 1978).

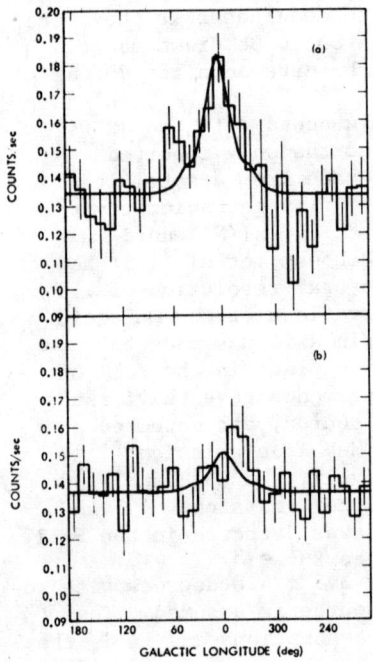

Figure 4

Net cosmic and background .511 MeV line flux as a function of galactic longitude measured by the gamma-ray spectrometer aboard the HEAO-3 spacecraft for the (a) Fall 1979 and (b) Spring 1980 galactic plane scans. The solid lines show the best-fit for a constant background level plus point source. (From Riegler, et al., 1981).

time variability that causes one to lean toward the more localized interpretation. Figure 5 shows the line spectrum with the peak position found to be 510.90 ± 0.25 MeV, with a width of 1.6 (+ 0.9, -1.6) keV. In the Spring, this emission measured (0.65 ± 0.27) x $10^{-3}$ photons/cm$^2$-s, a factor of three decrease. The probability that this decrease was a chance occurrence is 5 x $10^{-4}$. There was also an accompanying significant decrease in the low energy continuum of about 45 percent in the 60-300 keV energy band during the six month period between the observations[12]. Analysis of the data thus far yields an upper limit of 100 percent for the relative number of annihilations from the positronium states.

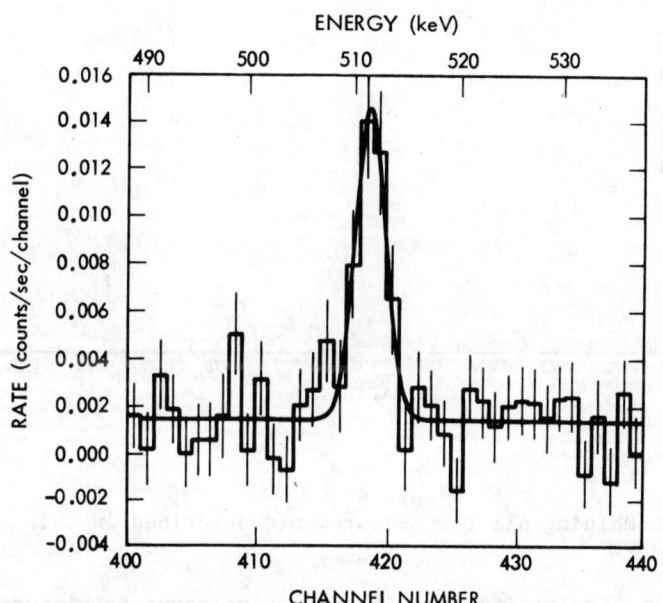

Figure 5

The net Galactic Center region line spectrum as measured by the JPL group during the Fall 1979 observations. The line is centered at 510.90 ± 0.25 keV, with a line width of 1.6 (+0.9, -1.6 keV). (From Riegler, et al., 1981).

## DISCUSSION AND CONCLUSIONS

Figure 6 shows the light curve defined by all the measurements described above. The observations, span a period of about 10 years, and average somewhat less than one per year. Even so, the light curve puts rather severe limitations on any source model advanced to explain it. If we make the assumption that what is observed is a single point source, then the weighted mean flux value is (1.27 ± .22) x $10^{-3}$ photons/cm$^2$-s, with chi-square per degree of freedom

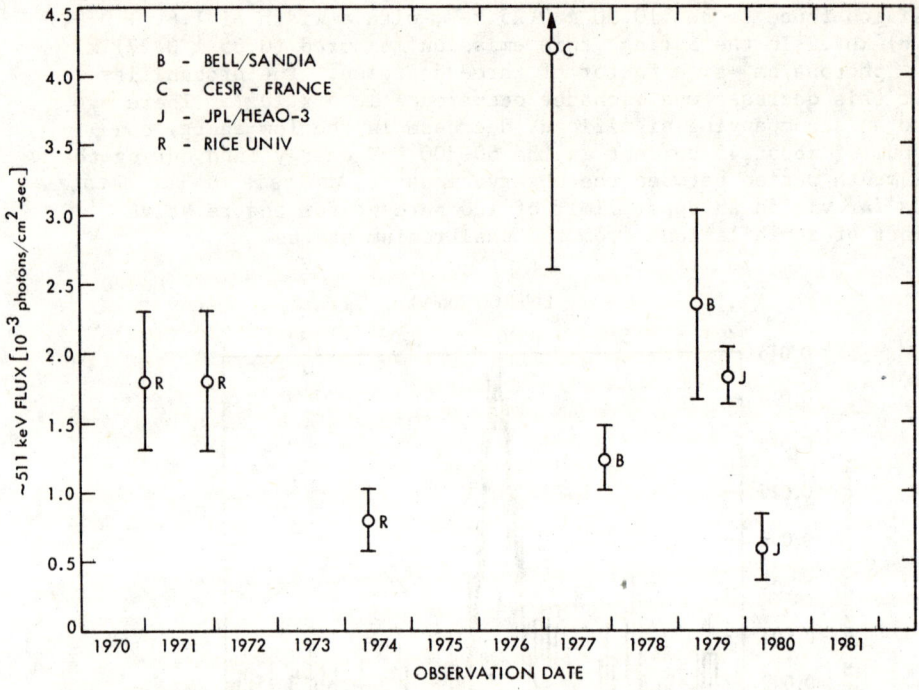

Figure 6
A light curve combining all the measurements described in this paper.

(d.o.f.) of 3.96, indicating that if this is an accurate picture of the source, it is probably a time varying one, and has been throughout the period over which the observations have taken place. Eliminating the last observation from the data set doesn't change this conclusion. Let's next assume that instead of being a single point source, the radiating region consists of an extended line source, which is constant, plus a point source. Let's further assume that the line source is of extent greater than fifty degrees, and that the last HEAO-3 observations was of the extended source only. Under these assumptions, the chi-square/d.o.f. is 2.60, still too high to be consistent with a constant point source. However, if we exclude the last HEAO-3 observation, the mean flux for the point source is found to be $(1.04 \pm .19) \times 10^{-3}$, with chi-square/d.o.f. of 1.80, consistent with a point source which was constant until sometime prior to the second HEAO-3 observation when its signal decreased significantly. From a purely statistical point of view, this is an acceptable model, and would be made more plausible by the detection of an extended source of .511 MeV line emission in the Galactic Center direction.

If the source is indeed at the Galactic Center, then it represents a luminosity in the .511 MeV line of about $10^{37}$ ergs/s, and a consequent annihilation rate on the order of $10^{43}$ per second. The narrowness of the line limits temperatures of the emitting region to about $10^5$ K. Time variability of less than about six months limits the extent of the annihilation region to less than $10^{18}$ cm in diameter. If the width, and peak positions are interpreted as limiting the gravitational red-shifts, then the mass within that radius must be less than $10^9$ $M_\odot$. Furthermore, if there is a mass of $10^6$ $M_\odot$ at the center, then the annihilations responresponsible for the observed line can take place no closer than $10^{14}$ cm. These circumstances, while not eliminating the cumulative contributions of multiple sources, do seem to point to a localized emission region for the variable source.

As reported elsewhere in these proceedings, subsequent balloon flight observations have tended to verify the time variability, and low state of the .511 MeV emission. What the source will do in the future and how its behavior will constrain acceptable models is very important to determine. A vigorous observing program is necessary to monitor the time variations to clearly establish their characteristic time scales, detect any possible peridicities, and answer many of the riddles posed by this enigmatic gamma-ray line source.

The author wishes to acknowledge Drs. W. A. Mahoney, and W. A. Wheaton for their helpful suggestions. This work was carried out under NASA contract NAS7-100 at the Jet Propulsion Laboratory, California Institute of Technology.

## REFERENCES

1. Johnson, W. N., and Haymes, R. C. 1973, Ap. J., 184, 103.
2. Kozlovsky, B., and Ramaty, R., 1974, Ap. J., 191, L43.
3. Fishman, G. J., and Clayton, D. D. 1972, Ap. J., 178, 337.
4. Leventhal, M. 1973, Ap. J., 183, L147.
5. Haymes, R. C., Walraven, G. D., Meegan, C. A., Hall, R. D., Djuth, F. T., and Shelton, D. M. 1975, Ap. J., 201, 593.
6. Albernhe, F., Leborgne, J. F., Vedrenne, G., Boclet, D., Durouchoux, P., and da Costa, J. M. 1981, Astr. Astrophys., 94, 214.
7. Leventhal, M., MacCallum, C. J., and Stang, P. D. 1978, Ap. J., 225, L11.
8. Leventhal, M., MacCallum, C. J., Huters, A. F., and Stang, P. D. 1980, Ap. J., 240, 338.
9. Mahoney, W. A., Ling, J. C., Jacobson, A. S., and Tapphorn, R. 1980, Nucl. Instr. Methods, 178, 363.
10. Jacobson, A. S. 1981, Ann. of N. Y. Acad. of Sci., 375, 330.
11. Riegler, G. R., Ling, J. C., Mahoney, W. A., Wheaton, W. A., Willett, J. B., Jacobson, A. S., and Prince, T. A., 1981, Ap. J., 248, L13.
12. Riegler, G. R., Ling, J. C., Mahoney, W. A., Wheaton, W. A., Jacobson, A. S., 1981, 17th ICRC, Paris.

## TIME VARIABLE POSITRON ANNIHILATION RADIATION FROM THE GALACTIC CENTER DIRECTION

M. Leventhal
Bell Laboratories, Murray Hill, NJ 07974

C. J. MacCallum
Sandia Laboratories, Albuquerque, NM 87185

### ABSTRACT

Our balloon-borne Ge gamma-ray telescope has recently been flown for a third time over Alice Springs, Australia to further study the galactic center 511 keV positron annihilation line. The source was found to be in an "off" or "low" state. A preliminary flux estimate of $0\pm4\times10^{-4}$ photons cm$^{-2}$s$^{-1}$ has been made. This result is discussed in the context of other observations and theoretical models.

### INTRODUCTION

On 1977 Nov. 11-12 the joint Bell Laboratories-Sandia Laboratories gamma-ray astronomy group flew a balloon-borne Ge telescope over Alice Springs, Australia in an attempt to detect spectral lines and continuum radiation from the galactic center (GC) direction. This flight resulted in the first unequivocal detection of a gamma-ray line from outside of the solar system[1,2] i.e., the 511 keV electron-positron annihilation line with a flux of $1.22\pm0.22\times10^{-3}$ photons s$^{-1}$ cm$^{-2}$. Some tantalizing but not conclusive evidence for a three-photon positronium continuum to the low energy side of 511 keV was also obtained. (The pioneering work of the Rice University group earlier in the decade had produced some suggestive but seemingly inconsistent evidence for a GC line near 500 keV.[3,4]) Since our 1977 flight, confirmation of the GC 511 keV line has come from several different balloon observations,[5,6,7] and most spectacularly in two separate observations by the JPL group employing the Ge spectrometer onboard the HEAO C space craft.[8] The JPL observations have also provided the first direct information concerning the spatial extent of the emitting region ($\lesssim 22°$) and strong evidence that the line flux is variable on a time scale of several months. We report in this paper the preliminary results of a third balloon flight over Alice Springs, Australia by the joint Bell Laboratories-Sandia Laboratories group.

### THE EXPERIMENT

The instrument was essentially the same as that described originally by Leventhal, MacCallum and Watts.[9] It is built around a single large, high-purity Ge detector operated at cryogenic temperature and surrounded by $\sim$ 200 kg of NaI in active anticoincidence.

The entrance aperature in the NaI shield defines the field of view, which is 15° FWHM at 511 keV. The energy resolution of the system has been improved from 3.2 keV to 2.4 keV FWHM at 511 keV by making changes in the electronics. Another important modification was the replacement of the central Ge detector with the largest volume (∼ 200 cm$^3$) high purity Ge detector ever made.

The flight took place on 1981 Nov. 21. Approximately 7 hours of GC data were obtained. As before, data were accumulated in alternate 15-20 minute target-background pair segments with the telescope maintained at the same zenith angle but rotated 180° in azimuth for the background measurements.

## RESULTS

While it is still too soon in our data analysis process to present fully massaged results, the essential conclusion may be gleaned from the raw data shown in Fig. 1. All workers in this field must contend with an instrumental 511 keV line from cosmic-ray generated positrons annihilating in the telescope and the atmosphere. In our previous flights the intensity of this line above the background continuum was observed to nearly double in the target spectrum versus the background spectrum indicating a source of annihilation radiation in the GC direction.[10] No such effect is present in the 1981 Nov. 21 data. It is apparent to the unaided eye that no statistically significant difference exists between the target and background 511 keV line intensity.

Before a final value for the 511 keV flux limit is derived a series of corrections and manipulations must be performed on the whole body of raw data.[1] These include corrections for (1) slight drifts in the calibration of the energy axis during the course of the observation, (2) variable instrumental dead time, and (3) variable atmospheric attenuation. Statistically weighted target-background pair difference rates will be formed and averaged over the observation. Finally the difference spectrum will be converted to a photon flux on a channel-by-channel basis by dividing by detector area and photo peak efficiency derived from laboratory measurements and Monte Carlo calculation.

To obtain a preliminary estimate of our limiting line flux, we have taken the data base readily available to us at this time (∼ 90% of the total) and performed by hand, approximations of the corrections and manipulations noted above. The resulting limiting 511 keV line flux is $0\pm4\times10^{-4}$ photons cm$^{-2}$ s$^{-1}$, where the quoted error is the usual one standard deviation. This result will surely change somewhat when the full blown computerized data analysis procedure is applied, but it is not expected to change the sense of the result.

## DISCUSSION

Table 1 summarizes the complete historical record of GC 511 keV line observations and places the present result in context. Clearly some of these results are of marginal statistical significance. However, if all of them are taken at face value, one might immediately conclude that there is an obvious time variability to the flux. Here one must be careful because a wide variety of instruments have been employed each with different potential systematic errors and different fields of view and energy resolution. If the source were constant and extended over the galactic plane, as some models[11] predict, an apparently larger flux should be reported by instruments with a larger field of view. Indeed just such an effect may be indicated by the CESR/CEN and UNH observations. To eliminate these possibilities we take a conservative point of view. Only successive measurements made by the same groups with essentially the same instruments should be compared when drawing conclusions concerning the temporal behavior of the flux. This leads to consideration of the satellite measurements by the JPL group and the balloon measurements by the Bell/Sandia group.

For two separate two-week long observing periods, starting on 1979 September 27 and 1980 March 4, the HEAO C Ge spectrometer scanned the galactic plane. Each time a positive 511 keV line flux was detected as shown in Table 1. Between the 1979 fall and 1980 spring observations the 511 keV flux decreased by $(1.20\pm0.35)\times10^{-3}$ photons $cm^{-2} s^{-1}$. The statistical likelihood of an upward or downward change in flux level by $3.5\sigma$ is $5.0\times10^{-4}$ for a normal distribution. Hence these workers concluded that they had observed a statistically significant change in flux over a period of $\sim$ 6 months. By making a light travel time argument they were then able to infer that the source contained at least one component that was $\lesssim 10^{18}$ cm in size. In fact they concluded that their observations were consistent with the emission coming entirely from a point source at the GC although they were not able to rule out a distributed component $\lesssim 22°$ in extent.

The new result reported here by the Bell/Sandia group indicates that the GC source has gone into an "off" or "low" state since our flights of 1977 and 1979. This result is entirely consistent with the HEAO C result and provides strong confirmation that the source intensity is time variable. It would appear that the source was going into a "low" state at the time of the 1980 March HEAO C observations and has remained there through 1981 November. At this time it seems safe to conclude that the time scale for change is at least as short as months indicating the presence of at least one relatively compact object.

The question of source extent seems less clear. If the new Bell/Sandia result means that the source has truly gone off, then it is hard to imagine how an extended source other than a collection of relatively compact objects can exist. The only attempt to directly map out the source extent was done by the JPL group. As noted above they concluded that their data excluded a distributed source component $\gtrsim 22°$ in extent. The CESR/CEN group claim their results are consistent with a distributed source as large as 60° in extent but do not rule out a point source. The other workers have not addressed this issue. It is tempting to ascribe the large flux values reported by the CESR/CEN and UNH groups to an extended source seen more fully by these large field of view instruments. However, the JPL and Bell/Sandia work seems quite secure and we conclude that any distributed contribution to the flux is $\lesssim 22°$ in extent and contributes only a small fraction of the total flux seen by these workers.

The next logical step is to relax restrictions on intercomparisons between different groups and to ascribe most of the variability seen in Table 1 to source fluctuations. The most restrictive of these comparisons is between the Bell/Sandia flight in 1977 November 11 and the UNH flight of 1977 November 21. If we exclude possible instrumental effects and the possibility of a distributed component, then the source intensity must have fluctuated by at least a factor of 2 in 10 days. This implies a source size of $\lesssim 10^{16}$cm. Probably it is premature to press the observational picture this far at this time. However, it is fair to conclude that all observations are consistent with a single relatively compact and time variable source at the GC.

Many possible physical origins for the GC 511 keV emission have been discussed in the literature. These include cosmic ray interactions in the interstellar medium, radioactive decay in supernova remnants, $e^+$-$e^-$ pair production in the strong magnetic fields of pulsars, electromagnetic and nuclear processes in the vicinity of a massive black hole, and the evaporation of primordial black holes. The most recent review of the theoretical situation is given by Ramaty and Lingenfelter.[11]

The observational constraints discussed earlier in this paper would now appear to be eliminating some of these possibilities. To account for the observations any model must allow $\sim 2 \times 10^{43}$ positrons to annihilate per second in a volume $\lesssim 10^{18}$cm across and in a natural way account for the time variable flux. In addition the narrow width of the 511 keV line[1,5,8] and the probable detection of the three-photon positronium continuum[1,7] imply that the annihilation site is one of relatively low density, partially ionized gas of temperature less than about $10^5$K.[12]

The fact that the emission is highly localized and variable would seem to eliminate cosmic ray interaction and radio pulsar models. Both these mechanisms should give rise to widely distributed and steady emission. Careful numerical considerations also dictate against such models. On the other hand, a galactic supernova or a massive rapidly rotating Kerr black hole still seem to be viable candidates. Black hole models have been given a boost by the recent infrared observations of Lacy et al.[13] These workers detected a group of high velocity, highly ionized clouds orbiting the GC which suggests the existence of a central massive object. The infrared emitting clouds also provide the proper kind of annihilation medium although a young supernova remnant will also do. Apparently no problem exists in generating the required variable positron flux in the accretion disc surrounding a rotating massive black hole.[11] Ultimately the question of positron genesis will be settled by further observation in both the gamma-ray and other regimes.

## ACKNOWLEDGMENTS

We wish to thank A. F. Huters and P. D. Stang for their invaluable managerial and engineering support; N. Corliss, L. Gillete, J. J. Lochtefeld and D. Sayers for superb technical support; and the men of the NCAR balloon launching team led by R. Kubara for an excellent flight.

The balloon, helium and some shipping expenses were provided by NASA.

The efforts of the Sandia personnel were supported by USDOE.

## REFERENCES

1. M. Leventhal, C. J. MacCallum and P. D. Stang, Ap. J. (Letters) 225, L11 (1978).
2. M. L. Cherry, E. L. Chupp, P. P. Dunphy, D. J. Forrest, and J. M. Ryan, Ap. J. 242, 1257 (1980).
3. W. N. Johnson and R. C. Haymes, Ap. J. 184, 103 (1973).
4. R. C. Haymes, G. D. Walraven, C. A. Meegan, R. D. Hall, F. T. Djuth and D. H. Shelton, Ap. J. 201, 593 (1975).
5. M. Leventhal, C. J. MacCallum, A. F. Huters, and P. D. Stang, Ap. J. 240, 338 (1980).
6. F. Albernhe, J. F. Leborgne, G. Vedrenne, D. Boclet, P. Durouchoux and J. M. da Costa, Astr. Ap. 94, 214 (1981).
7. B. M. Gardner, D. J. Forrest, P. P. Dunphy and E. L. Chupp, This volume (1982).
8. G. R. Riegler, J. C. Ling, W. A. Mahoney, W. A. Wheaton, J. B. Willett and A. S. Jacobson, Ap. J. (Letters) 248, L13 (1981).
9. M. Leventhal, C. J. MacCallum and A. C. Watts, Ap. J. 216, 491 (1977).
10. See Fig. 1 of Reference 1.
11. R. Ramaty and R. E. Lingenfelter, Phil Trans. R. Soc. Lond. A301, 671 (1981).
12. R. W. Bussard, R. Ramaty and R. J. Drachman, Ap. J. 228, 928 (1979).
13. J. H. Lacy, F. Baas, C. H. Townes, and T. R. Geballe, Ap. J. (Letters) 227, L17 (1979).

## FIGURE CAPTIONS

Fig. 1 Energy spectra in the vicinity of 511 keV for the sum of $\sim$ 90% of all GC target data (solid curve) and $\sim$ 90% of all GC background data (dashed curve). Individual raw data segments have been simply added together without making corrections of any kind. Only the instrumental 511 keV line is visible in both spectra.

TABLE I
HISTORY OF GC POSITRON ANNIHILATION
LINE EMISSION OBSERVATIONS

| Group | Field of View (FWHM) | Energy Resolution (keV) | Date | Line Energy (keV) | Line Flux 10⁻³ Photons cm⁻² s⁻¹ | Reference |
|---|---|---|---|---|---|---|
| Rice | 24° | 86 | 1971 Nov 20 | 476±24 | 1.8±0.5 | 3 |
| Rice | 13° | 60 | 1974 Apr 2 | 530±11 | 0.8±0.23 | 4 |
| CESR/CEN | 50° | 13 | 1977 Feb 14 | 511 | 4.18±1.56 | 6 |
| Bell/Sandia | 15° | 3.2 | 1977 Nov 11 | 510.7±0.5 | 1.22±0.22 | 1 |
| UNH | 100° | 40 | 1977 Nov 21 | 511 | 4.0±0.6 | this volume |
| Bell/Sandia | 15° | 3.2 | 1979 Apr 15 | 511 | 2.35±0.71 | 5 |
| JPL HEAO C | 35° | 2.7 | 1979 Oct | 510.90±.25 | 1.85±0.21 | 8 |
| JPL HEAO C | 35° | 5.7 | 1980 Mar | 510.1 | 0.65±0.27 | 8 |
| Goddard/CEN | 15° | 2.4 | 1981 Nov 20 | 511 | | this volume |
| Bell/Sandia | 15° | 2.4 | 1981 Nov 21 | 511 | 0.0±0.4 | this volume |

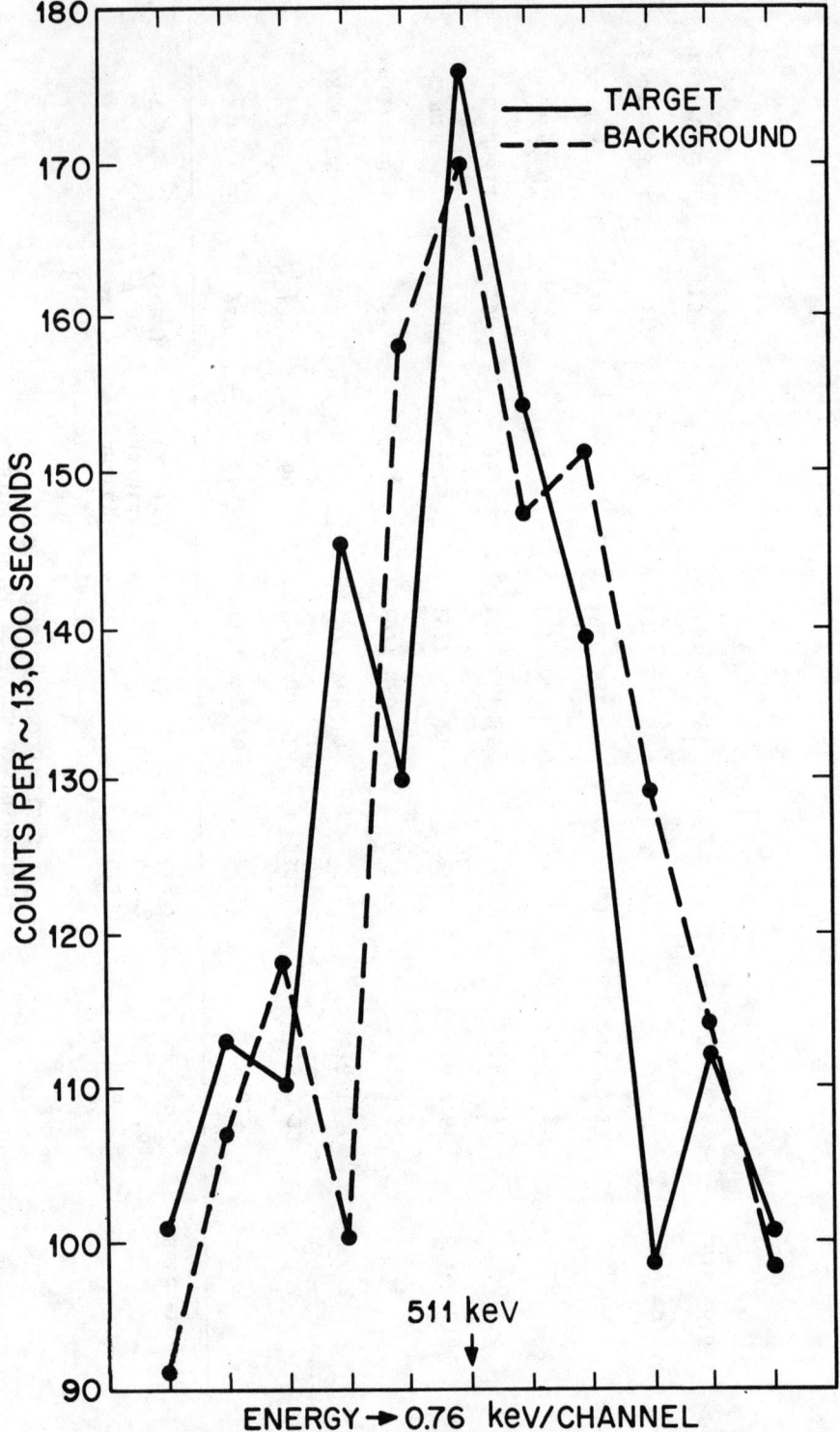

OBSERVATIONS OF THE GALACTIC CENTER WITH THE GSFC
LOW-ENERGY GAMMA-RAY SPECTROMETER: PRELIMINARY RESULTS

W. S. Paciesas*, T. L. Cline, B. J. Teegarden, and J. Tueller*
NASA/Goddard Space Flight Center, Greenbelt, MD 20771

P. Durouchoux, J. M. Hameury
Centre d'Etudes Nucleaires De Saclay, France

## ABSTRACT

The GSFC Low-Energy Gamma-Ray Spectrometer observed the galactic center on 20 November, 1981, during a balloon flight from Alice Springs, Australia. The positron annihilation line at 511 keV showed no excess over background when the galactic center was in the field-of-view; the inferred upper limit to the intensity of such a feature is $1.2 \times 10^{-3}$ photons/cm$^2$-s (95 percent confidence). Continuum emission was observed between 70 keV and 1 MeV.

## INTRODUCTION

A gamma-ray spectral feature was first observed to emanate from the region of the galactic center by a group from Rice University[1,2], but because of the relatively poor energy resolution of their instrument, neither the energy nor the shape of the feature was well established. The Bell/Sandia group[3] used a cooled germanium spectrometer to observe this region in 1977 and detected a narrow (FWHM ~ 3.2 keV) line feature centered on 511 keV. The JPL HEAO-3 instrument[4] performed two scans of the galactic center separated by 6 months and provided important new measurements of line energy, width, spatial extent, and variability. These, together with other recent reports[5,6] presented evidence for characteristics of the line which may be summarized as follows: 1) the energy is within 0.25 keV of 511 keV, so that any redshift must be small; 2) the width is less than ~ 2.5 keV; 3) the intensity appears to be variable; and 4) the continuum below 500 keV may show evidence for the 3-photon decay from the triplet state of positronium.

## OBSERVATIONS

In order to investigate the characteristics of the 511 keV line in further detail, we observed the galactic center with the GSFC Low-Energy Gamma-Ray Spectrometer during a balloon flight from Alice Springs, Australia, on 20 November, 1981. The instrument consists of an array of 3 coaxial germanium detectors with a total volume of 280 cm$^3$ which is collimated to 15° FWHM field-of-view by a thick cylindrical well of sodium iodide. The spectrometer has an energy resolution of 2.2 keV FWHM and an effective area of 11.8 cm$^2$ at 511

*Also the University of Maryland

keV. The telescope is mounted in a servo-controlled gondola using an alt-azimuth pointing system under microprocessor control. During the flight some problems were encountered with the latter system which resulted in our use of 2 modes of observation of the galactic center: a) a drift scan during which the instrument was pointed at the zenith (with no control of the azimuth) while the earth's rotation carried the galactic center (minimum zenith angle ~ 5.5°) through the field-of-view; and b) a pointed mode during which the instrument alternately tracked either the galactic center target or a point at the same elevation but 180° different in azimuth from the target (for background determination).

The data from each detector were analyzed separately for each observational mode. In the preliminary analysis of the drift scan it was assumed that all the emission emanates from the galactic center and that the collimator response function takes its zero-energy (i.e., geometrical) value. The former assumption breaks down if a significant fraction of the emission originates in a source located more than a few degrees away from the center but still within our field-of-view. This effect is probably most important below 100 keV where nearby sources may contribute significant emission. The latter assumption breaks down at high energies where the collimator is no longer opaque. Because of this, the drift scan data were <u>not</u> used above 511 keV. Minimum-$\chi^2$ fits were performed using the assumption of a constant source modulated by the detector response and superimposed upon a constant background.

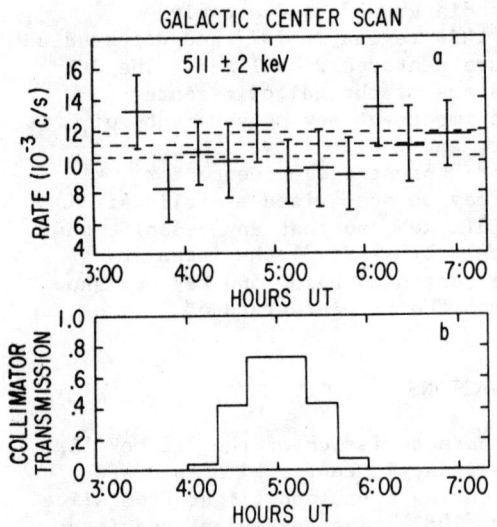

Figure 1. a) The counting rate in the 511 keV line as a function of time during the drift scan. The derived background rate and its ± 1σ errors are shown as dotted lines. b) The relative exposure to the Galactic Center during the drift scan.

The pointed mode data were analyzed by the usual method of taking the weighted mean of the target intervals and subtracting from it the weighted mean of the background intervals. Neither the drift scan data nor the pointing data showed any significant excess in the 511 keV line when the galactic center was in the field-of-view. For example, we show in Figure 1a the counting rate from all

3 detectors in a 4 keV wide bin centered on 511 keV as a function of time during the drift scan. The dotted lines show the instrumental background (± 1σ) obtained during the intervals when the source was not in the field-of-view. Figure 1b shows the fractional exposure of the galactic center during the scan. It is obvious that no significant source is present. When corrections for detector efficiency, instrumental deadtime, and absorption by intervening material are included, the resulting upper limit to the line intensity is $1.2 \times 10^{-3}$ photons/cm$^2$-s (95 percent confidence).

The continuum data were analyzed in a similar manner and are shown in Figure 2. For this case it is important to note that the detector efficiency corrections used only the full-energy peak response; i.e., the response matrix was assumed to be diagonal. This assumption is reasonable in our case because the Compton rejection efficiency of our NaI shield is high. However, because of this simplification we have not performed any spectral fitting on the data points and will only make qualitative assessments until a more complete treatment of the data is completed.

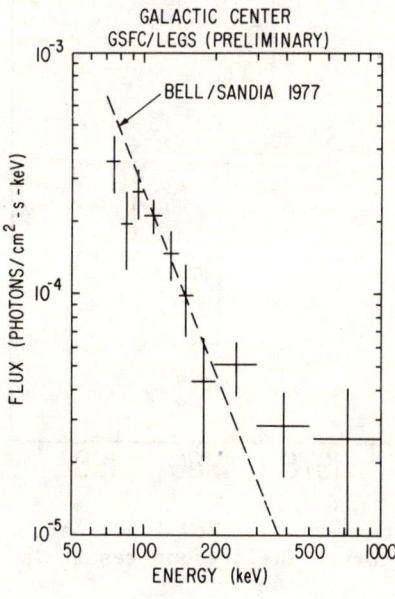

Figure 2. The continuum spectrum of the galactic center derived from preliminary analysis of LEGS data. Shown for comparison is the best-fitting power law obtained by Leventhal et al.[3]

## DISCUSSION

Figure 3 shows a time history of the intensity of the 511 keV feature in the galactic center. Our observation further strengthens the case for variability of the intensity of this feature. The present result is inconsistent at the 1σ level with all measurements except the 1974 Rice[2] point and the most recent (1980) HEAO-3 point[4]. The continuum spectrum, however, is not remarkably different from a number of previous observations. For example, we show in Figure 2 the actual power-law continuum measured by the Bell/Sandia group[3] during their 1977 flight. The agreement with the present data in the region where the systematic errors are smallest is remarkable. These authors also noted an excess in their continuum data just below 511 keV which they interpreted as evidence for positronium formation in the emitting region. The present data show a similar excess over the Bell/Sandia power-law with no corresponding 511 keV line. The apparent continuation of this excess to 1 MeV would argue against the positronium hypothesis, but we prefer to complete our drift scan analysis and model the detector response more carefully before placing too much emphasis on our 520-1000 keV data point, the significance of which is only $1.6\sigma$.

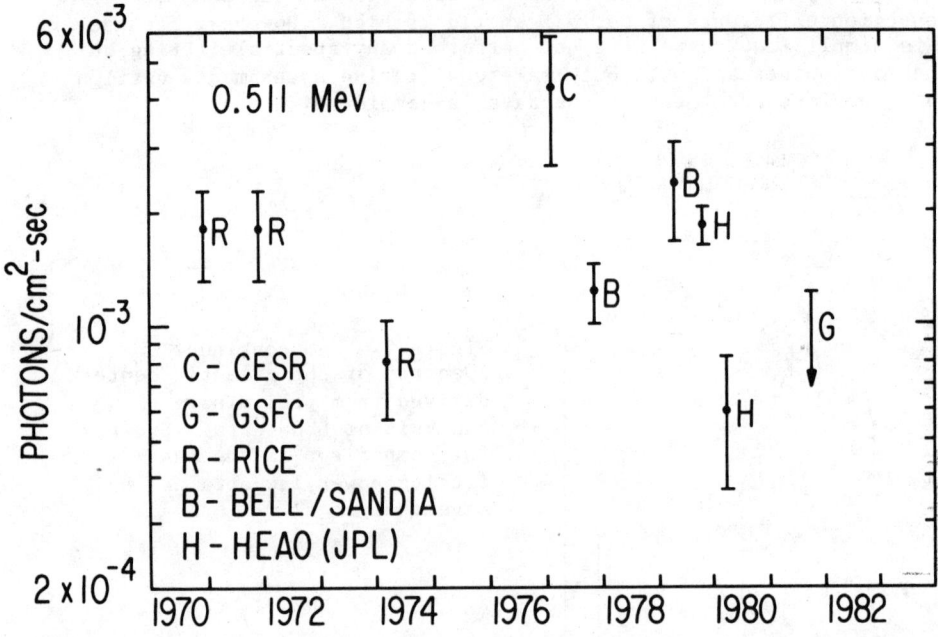

Figure 3. Time-history of the galactic center 0.511 MeV line. All positive detections are shown with $1\sigma$ errors. The present result is shown as a $2\sigma$ upper limit.

## SUMMARY

The galactic center was observed on 20 November, 1981, with a high resolution, cooled-germanium spectrometer. No significant emission was present in the region of the 511 keV annihilation line, the upper limit being $1.2 \times 10^{-3}$ photons/cm$^2$-s (95 percent confidence). This provides further evidence for the variability of this feature. Continuum flux was present in the 70-1000 keV band at a level remarkably similar both in intensity and spectral shape with that observed by the Bell/Sandia group[3] in 1977.

## REFERENCES

1. W. N. Johnson and R. C. Haymes, Ap. J. 184, 103 (1973).
2. R. C. Haymes, G. D. Walraven, C. A. Meegan, R. D. Hall, F. T. Djuth, and D. M. Shelton, Ap. J. 201, 593 (1975).
3. M. Leventhal, C. J. MacCallum, and P. D. Stang, Ap. J. 225, L11 (1978).
4. G. R. Riegler, J. C. Ling, W. A. Mahoney, W. A. Wheaton, J. B. Willett, and A. S. Jacobson, Ap. J. 248, L13 (1981).
5. M. Leventhal, C. J. MacCallum, A. F. Huters, and P. D. Stang, Ap. J. 240, 338 (1980).
6. F. Albernhe, J. F. Leborgne, G. Vedrenne, D. Boclet, P. Durouchoux, and J. M. da Costa, Astron. Astrophys. 94, 214 (1981).

NOTE: A more complete analysis of the data has been performed (Paciesas et al., 1982, paper submitted to Ap. J. Letters), and the results do not confirm the excess above the power law that was suggested in Figure 2. Furthermore, the inclusion of Compton energy loss effects in the response matrix results in ~ 20 percent lower flux for the data between 100 and 200 keV. Therefore, our data are inconclusive in regard to the positronium hypothesis.

# EMISSION IN THE 0.3 TO 1.0 MEV RANGE FROM THE GALACTIC CENTER REGION

B. M. Gardner, D. J. Forrest, P. P. Dunphy, and E. L. Chupp
University of New Hampshire, Durham, NH 03824

## ABSTRACT

The University of New Hampshire's Large Scintillator Gamma Ray Telescope observed the Galactic Center region on 21-23 November 1977, ten days after a similar observation by the Bell and Sandia Laboratories'[1] high resolution Ge gamma ray telescope. The UNH measurements are consistent with line emission near 511 keV with a flux that is at least twice as intense as the earlier measurement. The measurement requires a line intensity of $(4.0 \pm .6) \times 10^{-3}$ $\gamma$ cm$^{-2}$s$^{-1}$ and a flat spectral feature in the energy range 300 to 511 keV consistent with 70-80% $\beta^+$ annihilation from ground states of positronium.

## INTRODUCTION

An observation of the Galactic Center region was made with the Large Gamma Ray Telescope (LGT) flown from Alice Springs, Australia on 21-23 November 1977 by the University of New Hampshire's Gamma Ray Astronomy Group. The instrument is described in detail by Dunphy et al.[2] and basically consists of an actively shielded array of seven 7.6 x 7.6 cm NaI spectrometers. The energy resolution of the array is $\sim$ 40 keV FWHM at 511 keV and the absolute efficiency is based on preflight laboratory calibration and on the calculations of Berger and Seltzer[3]. This observation was made only ten days after that of Leventhal et al.[1] who suggest a model for emission from the Galactic Center region consisting of three spectral components; a power law continuum, a narrow line ($\leq$ 3.2 keV FWHM) at 0.511 MeV and a positronium continuum extending over energies $\leq$ 0.511 MeV. We test this model against the LGT measurements and compare the range of parameters with those reported by Leventhal et al.[1]

## ANALYSIS AND RESULTS

This analysis is limited to data accumulated from 2100 UT, 22 November to 0550 UT, 23 November. During this time altitude and detector pointing direction were constant and cutoff rigidity varied by no more than 8%, while the Galactic Center region rose from below the horizon to a position directly over the instrument, thereby drifting through the detector's field of view.

A direct comparison of count rates during this time is precluded by the changing rigidity. Therefore, to determine if our measurements indicate emission from the Galactic Center region, we accumulate 18 spectra, each for 1000 seconds, over the nine hour interval. Each spectrum is broken up into 14 discreet 0.35 MeV wide energy bands covering the range 0.21 to 5.10 MeV, and the fractional count rates ($r_i$) in each energy band are plotted versus time. Linear least squares fits done on each of the $r_i$ versus time indicate that at the 99% confidence level only $r_0$ (E = 0.21 to 0.55 MeV) is increasing in time. Figure 1 shows the time history of two fractional rates, $r_0$, and $r_1$ (0.56 to 0.90 MeV).

If the effect shown in Figure 1 is attributed to changing rigidity, then rigidity changes of the magnitude observed must be capable of producing dramatic spectral shape changes in the energy range 0.21 to 0.55 MeV. Such spectral changes are inconsistent with other observations made during this same flight as well as with observations made with this same instrument at 5 GeV (Palestine, Texas). Thus, we reject rigidity changes as causing spectral shape changes. However, the transmission coefficient for a source from the direction of the Galactic Center increased in a linear way from 0 to 0.7 over this same time interval. We therefore conclude that the effect can be attributed to emission from the Galactic Center region.

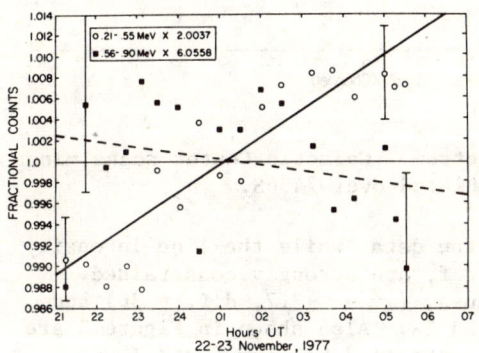

Fig. 1. Fractional rates versus time, solid line shows fits for $r_0$, dashed line shows fit for $r_1$.

Figure 2 shows the source energy loss spectrum obtained by subtracting a background spectrum, accumulated from 2105 to 2400 UT, 22 November 1977, from a source plus background spectrum, accumulated from 0215 to 0544 UT, 23 November 1977. A factor applied to the background spectrum to bring the two spectra to equal rigidities is estimated by forcing equal count rates in the energy range 0.6 to 1.0 MeV. Note that this assumes no significant emission in this energy range. Because this factor is not known exactly it must be considered a variable. A comparison of the background count rates between Palestine, Texas and Alice Springs, Australia indicates that this factor of 1.04 is reasonable.

The model suggested by Leventhal et al.[1] which includes a power law continuum, a positronium continuum and a line at 511 keV is tested using the source spectrum shown in Figure 2. In this case the width of the line response is fixed at 40 keV FWHM, the instrumental width, and the energy of the line is assumed to be 0.511 MeV. Free parameters are the intensity and power of the continuum, the intensity of the line ($I_0$) and the positronium fraction (f).[1] An extensive $\chi^2$ search reveals that the power law continuum parameters

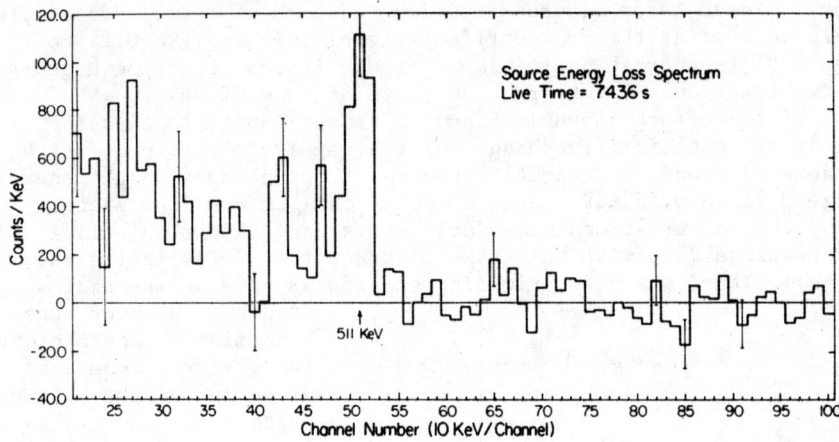

Fig. 2. Source energy loss spectrum. Galactic Center scans minus background scans. Counts accumulated over 7436S.

are virtually unconstrained by the data, while the line intensity and positronium fraction, $I_0$ and f, are strongly constrained. Figure 3 shows the best fit values, ($\chi^2$ = 47.7, d.f. = 36) and 99% confidence contours for f and $I_0$. Also shown in Figure 3 are the values and 99% confidence limits ($\pm$ 2.6$\sigma$) for f and $I_0$ obtained by Leventhal et al.[1] ten days earlier with a high resolution Ge detector.

## DISCUSSION

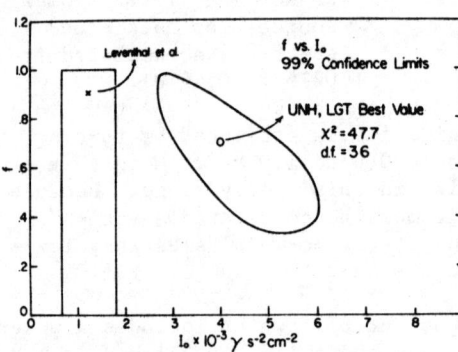

Fig. 3. Positronium fraction versus narrow line intensity. Contours show 99% confidence limits for this measurement and for measurement made ten days earlier.

The results of this observation are qualitatively consistent with measurements made ten days earlier, but are quantitatively different. Basic differences between the two experiments are energy resolution, angular response, and the method of separating the Galactic Center signal from the normal instrumental background. The Bell/Sandia[1] experimental values are ∿ 3 keV FWHM at 511 keV, with an angular FWHM of 15° and their background determination scheme emphasizes the differential flux only within their angular resolution around the Galactic Center region. The LGT energy resolution is 40 keV FWHM at 511 keV and the angular response of 100° FWHM allows contributions over a significant portion of the galactic disk as well as the center region.

We list below possible causes for the difference in line intensity as observed by the two experiments.

1.  The line intensity from the Galactic Center region changed over the ten days between measurements.

2.  The line at 511 keV consists of a narrow ($\sim$ 2 keV) component plus a broad ($\gtrsim$ 20 keV) component.

3.  The galactic disk is a significant source of line emission.

We do not feel that experimental measurements which would disprove either 2 or 3 have yet been performed.

## REFERENCES

1. M. Leventhal, C. J. MacCallum, and P. D. Stang, Astrophys. J. (Letters) 225, L11 (1978).
2. P. P. Dunphy, D. J. Forrest, E. L. Chupp, M. L. Cherry, and J. M. Ryan, Astrophys. J. 244, 1081 (1981).
3. H. J. Berger and S. M. Seltzer, NIM 104, 317 (1972).

## ON THE ORIGIN OF THE POSITRON ANNIHILATION RADIATION FROM THE DIRECTION OF THE GALACTIC CENTER

R.E. Lingenfelter
Center for Astrophysics & Space Sciences, C-011
University of California, La Jolla, CA 92093 USA

R. Ramaty
Laboratory for High Energy Astrophysics
NASA/Goddard Space Flight Center, Greenbelt, MD 20771 USA

### ABSTRACT

We review the observations of 0.511 MeV positron annihilation radiation from the direction of the Galactic Center, and discuss the implications of the line intensity, shape and time dependence both on the nature of the positron annihilation region and on the source of the annihilating positrons. We find that the positrons are most likely produced by photon-photon collisions in the vicinity of a compact object at the Galactic Center and that they annihilate within a distance of $\sim 10^{18}$ cm in a warm ($< 50,000$ K), partially ionized ($n_e/n > 0.1$), relatively dense ($\sim 10^5$ H/cm$^3$) gas.

### INTRODUCTION

Intense positron annihilation radiation at 0.511 MeV has been observed from the direction of the Galactic Center for over a decade. This emission was first seen in a series of balloon observations[1-3] with low resolution NaI detectors, starting in 1970. But it was not until 1977 that the annihilation line energy of 0.511 MeV was clearly identified with high resolution Ge detectors flown by Leventhal et al.[4,5] The latter observations also revealed that the line is very narrow (FWHM $\leqslant$ 3.2 keV) and that the continuum below 0.511 MeV could contain a significant contribution from positronium annihilation.

Recent Ge detector observations[6,7] on HEAO-3 have confirmed the existence of this very narrow line width (FWHM $<$ 2.5 keV) and have provided new information on the spatial extent of the emission region and on the time variability of the line intensity. These measurements showed that the line emitting region is smaller than the angular resolution of the detector (35°FWHM) and that within the observational uncertainty of $\pm 4°$, the direction of the source coincides with that of the Galactic Center. The HEAO-3 observations also showed that the 0.511 MeV line is time variable, decreasing by a factor of three in six months from an intensity of $(1.85 \pm .21) \times 10^{-3}$ photons/cm$^2$ sec in the fall of 1979 to $(0.65 \pm .27) \times 10^{-3}$ photons/cm$^2$ sec in the spring of 1980. This variability has been recently confirmed by balloon-borne, Ge detector observations made in the fall of 1981 by Leventhal and MacCallum and Teegarden et al., as discussed in these proceedings. Furthermore, NaI detector observations in the fall of 1977 (Gardner et al., in these proceedings) could indicate a variation on a time scale as short as 10 days.

The hard X-ray ($>$ 70 keV) intensity from the direction of the Galactic Center was already known[5] to be variable and the observed intensity at these energies, where the continuum from positronium annihilation is negligible, shows variations that are qualitatively similar to those of the annihilation line (Figure 1), suggesting that much of the hard X-ray continuum emission may also come from the same source.

These measurements of the 0.511 MeV line intensity, width and variability, place strong constraints on the nature of both the positron annihilation region and the positron source. In this paper we consider first the constraints on the annihilation region, then discuss possible positron sources in the light of further constraints imposed by the 0.511 MeV observations and those at other wavelengths.

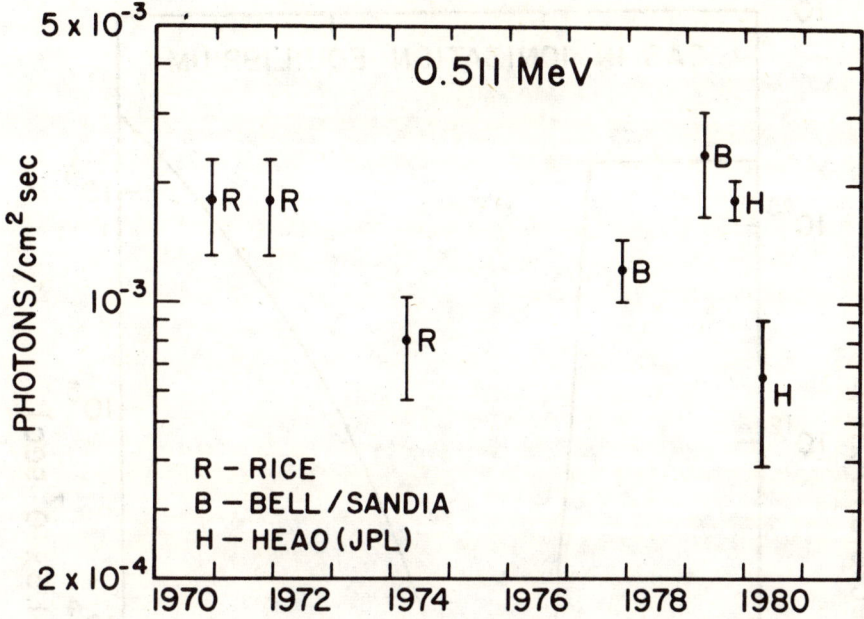

Figure 1. Annihilation line intensity from the direction of the Galactic Center, measured by groups at Rice University[1-3], Bell-Sandia[4-5] and the Jet Propulsion Laboratory[6-7]

## THE POSITRON ANNIHILATION REGION

The nature of the positron annihilation region is constrained by the line width and the intensity variations. The observed[4,7] line width of less than 2.5 keV requires[8] that the positions annihilate in a gas that is at least partially ionized (i.e. $n_e > 0.1\, n$). If the gas were neutral, the line width would be larger than observed because it would be Doppler broadened, not by the thermal motion of the gas, but by the velocity of energetic positrons forming positronium in flight by charge exchange with neutral hydrogen. In a partially ionized gas, however, positrons lose energy to the plasma fast enough that the positrons thermalize before they annihilate or form positronium. The line width thus reflects the temperature of the medium, so that the observations require a temperature $\leqslant 5 \times 10^4$ K.

The observed[4,7] line width further limits any velocities of rotation, expansion or random motion to $< 700$ km/sec. In addition, the observed[7] line center of $510.90 \pm 0.25$ keV, corresponding to a line shift of $-0.35$ keV $< \Delta E < +0.15$ keV, implies a bulk velocity along the line of sight of $-90$ km/sec $< V < +200$ km/sec, and a gravitational red shift $z < 7 \times 10^{-4}$.

Study[4] of the continuum emission observed at energies less than 0.511 MeV suggests that the bulk ($\sim 90\%$) of the annihilations take place via positronium. This ratio of positronium annihilation to direct annihilation implies a yield of 0.511 MeV photons per positron annihilation of $\sim 0.65$. Direct annihilation would give a ratio of 2, while positronium annihilation alone would give a ratio of 0.5.

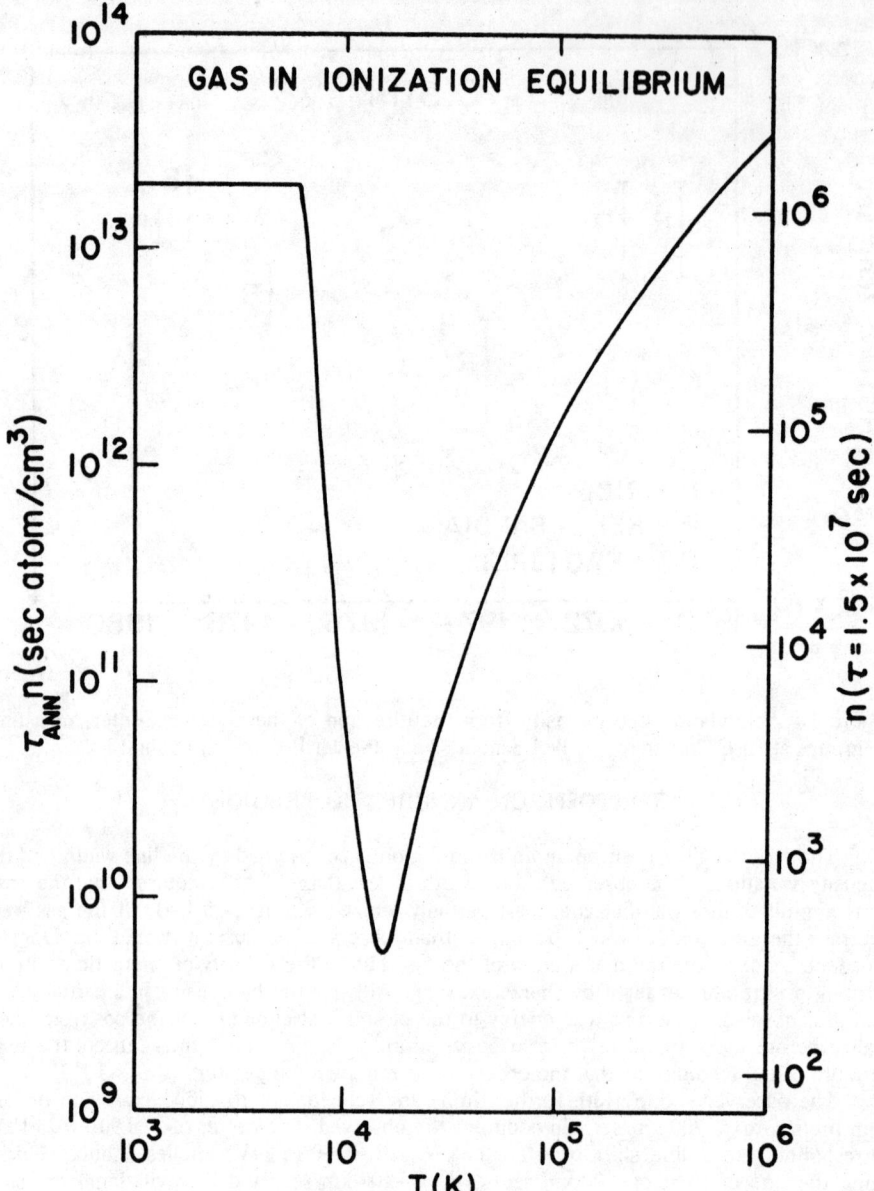

Figure 2. Thermal positron annihilation time as a function of temperature in a gas of solar composition in ionization equilibrium.

This ratio of positronium annihilation to direct annihilation is also consistent with that expected from the constraints on the temperature and degree of ionization imposed by the observed line width. The relative rates of direct annihilation on neutral hydrogen and free electrons and of annihilation via positronium, formed by charge-exchange with neutral hydrogen and radiative combination with free electrons, have been calculated by Bussard, Ramaty, and Drachman[8]. These calculations show that, if T < 50,000 K and $n_e/n$ > 0.1, more than 70% of the positrons form positronium and less than 30% annihilate directly. The positronium is not broken up before it annihilates[9], as long as the gas density is < $10^{15}$ H/cm$^3$.

Lastly, but most importantly, the observed intensity variation strongly constrains the size of the annihilation region and the density of matter in it. The size of the region should obviously not exceed a lightyear ($10^{18}$ cm) and either the density should be high enough that positrons can slow down and annihilate in less than half a year or the region should be small enough that it can be eclipsed in that time.

We consider first the limiting density needed to annihilate the positrons in < 1 year. The annihilation time[8] for thermal positrons is inversely proportional to the gas density and its dependence on temperature is shown in Figure 2 for a gas of solar composition in ionization equilibrium[10]. As can be seen from this figure, an annihilation time $\leq 1.5 \times 10^7$ sec requires a density $n \geq 400$ H/cm$^3$, even at the optimum temperature of $\sim 1.5 \times 10^4$ K. Since the short annihilation times result from annihilation via positronium formed by charge-exchange with neutral hydrogen, any photoionization of the gas would reduce the neutral fraction, increase $\tau_{ann}n$ and hence increase the minimum density required for annihilation in less than half a year.

The slowing down time sets an even more stringent limit on the density. Most production processes would be expected to generate positrons with initial kinetic energies on the order of $m_e c^2$ or greater. The limits on the observed annihilation line width require that the positrons slow down to energies corresponding to temperatures $\leq 5 \times 10^4$ K. Calculations (J. McKinley personal communication) of slowing down by Coulomb scattering and plasma wave interactions show that gas densities $\geq 10^5$ H/cm$^3$ are required for slowing down times of $\leq 1.5 \times 10^7$ sec. This limit could be relaxed somewhat if Compton scattering, synchrotron emission, or other energy loss processes were important.

Thus we see that the observed width and time variations of the annihilation radiation require that the annihilation region have a temperature of $\leq 5 \times 10^4$ K, a density of at least 400 H/cm$^3$ and probably $> 10^5$ H/cm$^3$, a degree of ionization $n_e/n \geq 0.1$ and dimensions of < $10^{18}$ cm. Such conditions can be found in stellar atmospheres and in both the peculiar warm clouds[17] and the other compact IR sources[19] observed within the central parsec of the galaxy.

On the other hand the rapid decrease in the annihilation radiation luminosity could in principle result from obscuration of the source, such as an eclipse of a stellar source by a close binary companion or an eclipse of a compact source at the Galactic Center by one of the clouds or giant stars around it. In either case an eclipse in less than half a year by an object with an orbital velocity on the order of 100 km/sec would require the diameter of the annihilation region to be $\leq 10^{14}$ cm.

## THE POSITRON SOURCE

The nature of the positron source is also strongly constrained by the observed variations of the 0.511 MeV intensity and by observations at other wavelengths. The decrease by a factor of three in the line intensity in six months clearly excludes any of the multiple, extended sources, such as pulsars[11], supernovae[12], or primordial black holes[13], previously proposed. Instead, it essentially requires a single compact (< $10^{18}$ cm) source which is apparently located either at or close to the Galactic Center and which either is inherently variable or is eclipsed on time scales of six months or less.

The observed line intensity (Figure 1) of $\sim 2 \times 10^{-3}$ photons/cm² sec requires at the distance of the Galactic Center ($\sim 10$ Kpc) a positron annihilation rate of $\sim 4 \times 10^{43}$ e⁺/sec, if $\sim 90\%$ of the positrons annihilate via positronium as implied by both the theoretical calculations and the continuum measurements[4]. This rate corresponds to a luminosity of $\sim 6 \times 10^{37}$ erg/sec in annihilation radiation, including both line and three-photon continuum emission. If these positrons are produced as electron-positron pairs with kinetic energies equal to their rest mass energy then the total required power is $\sim 1.2 \times 10^{38}$ erg/sec.

Comparing this annihilation radiation luminosity with limits on the luminosity of the Galactic Center region at other wavelengths (Figure 3), we find that there are two other important constraints on the positron source. These are the surprisingly large (nearly 50%)

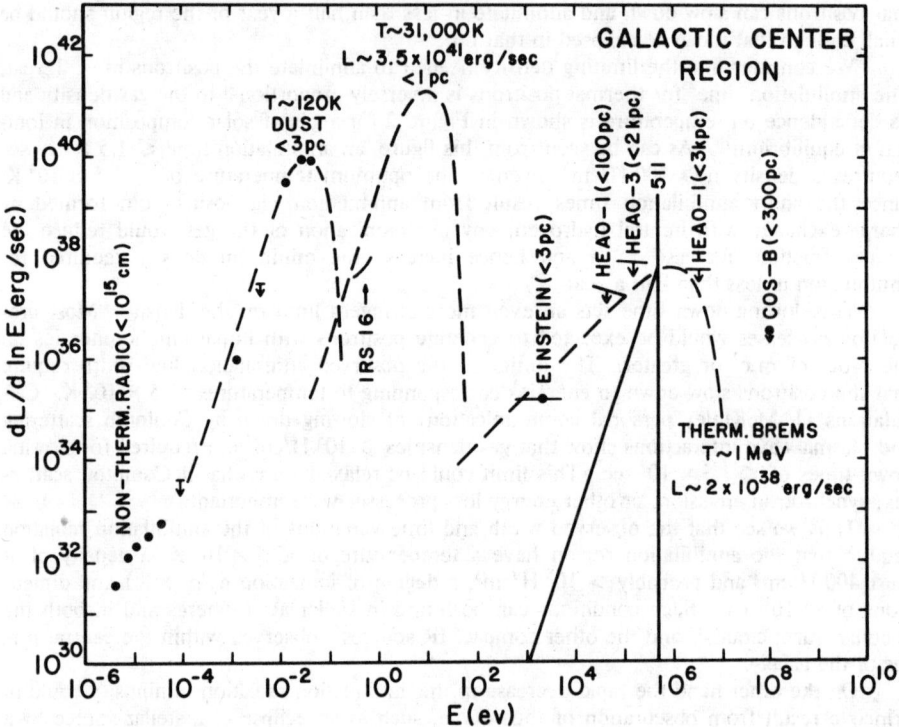

Figure 3. Limits on the luminosity per unit ln E as a function of photon energy from the region around the Galactic Center. Data are shown for the compact ($< 10^{15}$ cm) nonthermal radio source[19], the $\sim$ 3pc dust ring[16], the nonthermal infrared source IRS 16 (Ref. 18), the soft X-ray emission ($< 3$pc) from the EINSTEIN satellite measurements[20], the hard X-ray emission from HEAO-1 (Matteson in these proceedings) and HEAO-3 (Ref. 15), the 511 KeV line and positronium continuum[4,7], and the gamma ray emission from HEAO-1 (Ref. 14) and COS-B (Ref. 27). Also shown as dashed curves are the blackbody luminosity at $\sim$ 31,000 K required[17] to account for ionization in the warm IR clouds within $< 1$pc of the Galactic Center, a blackbody luminosity at 120 K as inferred[16] from the far infrared observations of the $\sim$ 3pc dust ring, and the maximum thermal bremsstrahlung luminosity at $\sim 1$ MeV, consistent with X- and gamma-ray observations.

ratio of the annihilation radiation luminosity to the hard X-ray and gamma-ray continuum luminosity and, if the source is, in fact, within the central parsec of the galaxy, the relatively large lower limit ($> 10^{-4}$) on the ratio of the annihilation radiation luminosity to the estimated bolometric luminosity.

The hard X-ray and gamma-ray continuum observed[4,14] from a 15° or more field of view around the Galactic Center in 1977 sets an upper limit on the continuum luminosity ($> 70$ keV) of any single source at the distance of the Galactic Center of $< 3 \times 10^{38}$ erg/sec. Thus the power required to produce the observed annihilation radiation, given above, is close to 50% of the total power radiated above 70 keV - a very strong constraint on any model. Comparable ratios of annihilation to hard X-ray continuum luminosity can be obtained from other observations[3,15] of the emission from the direction of the Galactic Center.

If the annihilation line source is, in fact, within the central few parsecs of the Galaxy, an important lower limit can also be set on the ratio of the annihilation radiation luminosity to the estimated bolometric luminosity. The most intense emission observed from this region is in the far infrared $\sim 10^{40}$ erg/sec, apparently thermal emission from a toroidal dust belt of $\sim$ 3pc diameter surrounding the dynamic center of the Galaxy. Gatley and Becklin[16], however, calculated that a larger luminosity of $4 \times 10^{40}$ to $1.2 \times 10^{41}$ erg/sec in the soft ultraviolet is needed to heat this dust. A similar central luminosity is inferred by Lacy et al.[17] from observations of infrared fine structure lines from the warm ionized gas clouds within the central parsec of the Galaxy. They calculate that, if the clouds are heated by a single central source, it must have a bolometric luminosity of $\sim 3.5 \times 10^{41}$ erg/sec and an effective temperature $\leqslant 31,000$ K. However, since there are at least $3 \times 10^6$ $M_\odot$ of stars within this region which are most likely the dominant contributors to the heating, this is necessarily only an upper limit on the bolometric luminosity of a central source. Comparing this luminosity with that of the annihilation radiation, we see that the annihilation radiation must carry more than $10^{-4}$ of the entire bolometric luminosity of the positron producing source, if it is within the central parsec.

Lastly we note that the luminosity of the annihilation radiation is comparable to the most intense emission directly observed so far at any other wavelength from the compact source or sources at the Galactic Center, i.e. that in the infrared band (1.65 to 4.8 $\mu$m) observed[18] from IRS 16. Furthermore it exceeds by a factor of $\sim 10^4$ the radio (1.35 to 31 cm) luminosity observed[19] from the nonthermal source at the Galactic Center and it is $\sim 4 \times 10^2$ times as intense as the soft X-ray (0.5 to 4.5 keV) luminosity of $1.5 \times 10^{35}$ erg/sec measured[20] by the Einstein Observatory from the central few parsecs of the Galaxy.

## POSITRON PRODUCTION PROCESSES

We now consider what constraints may be placed on the possible positron production processes in a source at the Galactic Center by the observed emission at various energies.

Positrons can be produced directly and indirectly by a number of processes:

$e^+$ decay of $\pi^+$ mesons produced in high energy nuclear interactions and matter-antimatter annihilation;

$e^+$ decay of radionuclei produced in low energy nuclear reactions;

$e^\pm$ pair production in electron-electron and electron-nucleus interactions;

$e^\pm$ pair production in photon-electron and photon-nucleus interactions;

$e^\pm$ pair production in photon-photon interactions; and

$e^\pm$ pair production in interactions of both photons and electrons with intense magnetic fields.

The studies that we have made so far do not allow us to uniquely determine what positron production process is responsible for the annihilating positrons. Nonetheless if we assume that the time scale for production of the positrons is comparable to the time scale for annihilation, that the positron production region is optically thin to gamma rays and that gamma radiation from it is isotropic, then we can effectively exclude all but one of the above processes - pair production in photon-photon interactions.

But if we relax those assumptions allowing for the possibility that the positrons might be produced in a short impulsive burst or bursts at any time when the Galactic Center was not being observed in X- and gamma-rays, that the positron production region is optically thick to gamma rays, or that gamma radiation from that region is strongly beamed in a direction out of the line of sight between us and the source, then the constraints implied by the observations are likewise relaxed and only a few processes can effectively be excluded.

What follows then is a preliminary discussion of the observational constraints that may be imposed on the various possible positron production processes that might be responsible for the annihilation radiation observed from the direction of the Galactic Center. We start with those processes which seem to be the least likely.

Electron-positron pair production by photon interactions with intense magnetic fields[21] does not appear to be adequate to account for the annihilating positrons in the direction of the Galactic Center. Sturrock[22] has suggested that in the intense magnetic fields ($\geqslant 10^{12}$ G) expected near the polar caps of pulsars an $e^{\pm}$ pair cascade may develop from $e^{\pm}$ curvature radiation producing gamma rays which in turn produce further $e^{\pm}$ pairs by interaction with the magnetic field. But the calculated[11] rate of positron production by a single pulsar in this model[22] does not exceed $\sim 10^{41}$ positrons/sec, so that such a pulsar would have to be closer than 1 kpc. Moreover, other pulsar models[23,24] give much lower positron production rates of $10^{35}$ to $10^{36}$ positrons/sec.

Positrons from the decay of $\pi^+$ mesons, produced either by relativistic particle interactions or by matter-antimatter annihilation, also do not appear to be an important source of the observed annihilation radiation. Since the production of $\pi^+$ mesons in both of these processes is accompanied by the production of $\pi^\circ$ mesons which decay into high energy gamma rays, the gamma ray observations[25-27] (> 100 MeV) by SAS-2 and COS-B during the last decade can limit the contribution of $\pi^+$ decay positrons to the observed 0.511 MeV emission. Taking the ratio of annihilating positrons to 0.511 MeV photons to be $\sim 0.65$, as discussed above, the ratio of $\pi^+$ to $\pi^\circ$ production to be $\sim 1$, and the ratio of > 100 MeV photons to $\pi^\circ$ mesons to be $\sim 1$, the steady state ratio of 0.511 MeV photons from annihilation of $\pi^+$ decay positrons to > 100 MeV photons from $\pi^\circ$ decay is $\sim 1.5$. The observed[27] flux of $1.8 \times 10^{-6}$ photons (> 100 MeV)/cm$^2$ sec from a point source of < 2° radius at the Galactic Center limits the expected flux at 0.511 MeV from annihilation of $\pi^\pm$ decay positrons to only $\sim 10^{-3}$ of the observed 0.511 MeV flux. Even taking the high energy gamma ray flux observed[25,26] from the full 15° field of view of the 0.511 MeV line detectors, the contribution of $\pi^+$ decay positrons would be only $\sim 10^{-2}$ of that required to account for the observed flux. Thus, unless the pions were produced in exceedingly intense bursts which escaped detection by the SAS-2 and COS-B detectors, $\pi^+$ decay can be excluded as the source of the annihilating positrons from the direction of the Galactic Center.

Positrons from radioactive decay of $^{56}$Ni → $^{56}$Co → $^{56}$Fe or other nuclei produced in explosive nucleosynthesis[28] do not seem to be a suitable source either. Even though a single supernova occurring close to the Galactic Center could release[29] sufficient ($\sim 10^{52}$) positrons to account for all of the annihilation radiation observed over the last decade, there are serious difficulties with such a suggestion. First, the gamma ray observations[1-7,30] exclude any supernova occurring even within ±15° of the Galactic Center at any time after about 1965, for if one had occurred the emission in the 0.847 MeV line from $^{56}$Co decay would have greatly exceeded the observed intensity at that energy. Then if the supernova occurred sometime prior to about 1965 there does not appear to be any obvious way that such an event could produce the time variations in the 0.511 MeV flux observed over the last decade.

The question of whether sufficient positrons might be produced from decay of radionuclei made in thermonuclear burning in some other process of catastrophic stellar disruption or collision close to the Galactic Center remains to be studied.

Positrons can result from low energy (< 100 MeV/nucleon) nuclear interactions either by decay of spallation nuclei, e.g. $^{11}C(e^+)^{11}B$, or by deexcitation of excited nuclear levels, e.g. $^{16}O^{*6.05}(e^\pm)^{16}O$. Calculations[31-33] of the positron production rates by these processes have been made for a wide range of assumed energetic particle spectra and compositions, and complementary calculations[34] have been made of the accompanying gamma ray deexcitation line emission. These studies show that over a wide range of particle spectra and composition the intensity of the 4.44 MeV line from $^{12}C^*$ deexcitation should be nearly equal to that of the 0.511 MeV line from positron annihilation. A very broad line at about 4.6 MeV with an intensity equal to that of the 0.511 MeV line from the Galactic Center was reported[3] from the same direction during the 1974 observation. But only an upper limit of < 20% of the 0.511 MeV line intensity could be set[12] from the observational upper limits on the 4.4 MeV line in 1977 obtained[14] by the HEAO-1 gamma ray detectors. In view of the apparent time variations in the positron production and annihilation rates, however, the gamma ray line observations can neither confirm nor exclude energetic particle interaction as the source of the annihilating positrons. But the ratio of the observed annihilation radiation luminosity to the limiting bolometric luminosity of the central parsec, discussed above, does severely constrain possible positron production by such interactions. In particular, this ratio ($> 10^{-4}$) would exceed that expected[31-32] for nuclear interactions by at least an order of magnitude if the bulk of the energy lost by the energetic particles is eventually radiated away. Therefore, if such interactions are the source of the positrons, then the bulk of the energetic particle energy loss must be carried away by lower energy particles either into the interstellar medium where it is radiated from a much larger volume ($>> 1pc^3$) or into a black hole[32].

Finally electron-positron pair production by all combinations of interactions between photons, electrons and ions can be divided into two general situations: interactions in which the photon, electron and ion energies are all comparable, and interactions of high energy ($>> m_ec^2$) photons and electrons with much cooler ambient photons and gas. We consider the latter situation first.

A pair production cascade by interactions between ambient photons and very high energy beamed electrons, positrons and gamma rays generated by dynamo action around a massive accreting black hole in galactic nuclei has been suggested by both Blandford[35] and Lovelace et al.[36] as a source of the energetic particles responsible for the powerful radio emission of active galaxies. We have proposed[37] that this process might also be responsible for the annihilation line from the Galactic Center. Electron-positron pair production by such high energy gamma rays with initial energies, $E^o$, interacting with ambient photons of mean energy $\epsilon$ will proceed as long as the gamma ray energy exceeds the pair production threshold of $\sim (m_ec^2)^2/\epsilon$. But since pair production stops once the gamma ray energy falls below the threshold, the maximum fraction of the initial energy available for annihilation radiation is $\sim \epsilon/m_ec^2$. Within the central parsec of the galaxy the most numerous ambient photons are most likely those in the $\sim$ 30,000 K radiation, as discussed above in connection with the bolometric luminosity. For such photons with $\epsilon \sim$ 3 ev, the maximum fraction of the high energy luminosity that could go into annihilation radiation is only $\leqslant 10^{-5}$. Thus, if this were the source of the annihilation positrons, the annihilation luminosity of $\sim 6 \times 10^{37}$ erg/sec would require a luminosity of $> 6 \times 10^{42}$ erg/sec in high energy photons. Since this exceeds by at least four orders of magnitude the observed upper limits on the luminosity at energies $> m_ec^2$, such emission would have to be highly collimated and beamed out of the line of sight. Even then this bolometric luminosity would be comparable to that of a Seyfert galaxy, such as NGC 4151.

The required high energy photon luminosity could be reduced significantly, however, if the ambient gas density is greater than about $\alpha^{-1}$ (i.e. 137) times the ambient photon density. Then more electron-positron pairs would be produced by a cascade interaction with gas than by that with ambient photons. In this case high energy gamma rays interacting with the gas will continue to make pairs until the pair energies fall below a critical energy $E_c$ at which the rate of energy loss by bremsstrahlung falls below that by Coulomb scattering. Thus the maximum fraction of the initial energy available for annihilation radiation is $\sim m_e c^2/E_c$. In a gas of roughly solar composition $E_c \sim 400$ MeV, so that the maximum fraction of the initial energy available for annihilation radiation is $\sim 10^{-3}$. In this case the observed annihilation radiation would require a high energy photon luminosity of $> 6 \times 10^{40}$ erg/sec which could still require beaming out of the line of sight but it would not need to be as highly collimated.

For such a cascade in the gas to be significant, however, requires that the ambient electron density be at least $\sim 10^2$ times the ambient photon density, simply the ratio of the pair production cross sections in photon-photon and photon-electron interactions. The estimated[17] luminosity of $\sim 3 \times 10^{41}$ erg/sec in $\sim 31,000$ K photons within the central parsec of the galaxy requires an average photon density of at least $\sim 3 \times 10^4$ photons/cm$^3$. Thus in order for a gas cascade to be important the gas density in the beam must be $> 3 \times 10^6$ H/cm$^3$. This is much greater than the average gas density[38] within the central parsec, but it is not inconsistent with the apparent densities in the warm ionized gas clouds observed[17] in the infrared. The passage of clouds into and out of such a beam could also produce time variations in the positron production rate.

Much more efficient electron-positron pair production, however, could occur, either in high energy photon beams by interactions of beam photons with one another (M. Burns, personal communication, 1982) or in a hot plasma where the electrons, ions and photons all have energies of the order of $\sim m_e c^2$. The relative importance of pair production by particle-particle, particle-photon, and photon-photon interactions again depend on the ratio of the photon to particle density. We consider the two extreme cases where this density ratio is either $\ll \alpha$ or $\geqslant \alpha$, the fine structure constant.

If the ratio of the photon density to that of the particles is much less than $\alpha$, the principal pair production process is particle-particle interactions, and the maximum ratio of the pair production rate to the bremsstrahlung emission rate is the maximum ratio of the average cross sections, approximately $\sigma_{pp}/\sigma_b \sim \alpha^2 \sigma_T / \alpha \sigma_T \sim \alpha$. Assuming that the average energy of the pairs is of the same order as the average energy of the bremsstrahlung photons, then the maximum fraction of the radiated energy available for annihilation radiation is of the order $\sim 10^{-2}$. Since this is still less than the observed lower limit of $\sim 10^{-1}$ on the ratio of annihilation radiation luminosity to the total luminosity at energies $\geqslant m_e c^2$, even this process could only account for the observed annihilation radiation if the pair production occurred in more intense, but unobserved bursts.

If, however, the ratio of the photon to particle density is on the order of $\alpha$ or greater, then photon-photon interactions would become the principal pair production process and the ratio of pair luminosity to continuum luminosity at energies $\geqslant m_e c^2$ could be consistent with the value of $\geqslant 10^{-1}$ observed in the direction of the Galactic Center.

The most efficient pair production occurs at photon energies close to $m_e c^2$. To first order the pair production rate in a spherical source of diameter d may be approximated by $Q \sim \frac{1}{2} n_\gamma^2 <\sigma c> d^3$, where $<\sigma c>$ is the average pair production cross section times the velocity of light (equal[39] to $\sim 3 \times 10^{-15}$ cm$^3$/sec for black body photons of temperature $\sim m_e c^2$) and $n_\gamma$ is the photon number density. Assuming that the source is optically thin this density can be related to the luminosity $\geqslant m_e c^2$ by $L \sim m_e c^2 n_\gamma c \pi d^2$. Thus for a given limiting luminosity at energies $\geqslant m_e c^2$, the positron production rate depends only on the size, such that $d \sim 3 \times 10^{-25} L^2/Q$ in cm. For the observed[14] limiting luminosity $< 2 \times 10^{38}$ erg/sec at energies $\geqslant m_e c^2$ and a positron production rate equal to the annihilation rate of $\sim 4 \times 10^{43}$ e$^\pm$/sec, we find that the diameter of the positron source d must be $\leqslant 3 \times 10^8$ cm. The region in which the escaping positrons annihilate, of course, may be much larger.

This production process thus requires an exceedingly compact source. A blackhole with such a Schwarzschild diameter must have a mass $\leqslant 500$ M$_\odot$ which is much smaller than the $10^6$ to $10^7$ M$_\odot$ blackholes that have been suggested[17,40] at the Galactic Center. Yet such a small size would be consistent with arguments by Ozernoy[41] that the Galactic Center cannot contain a blackhole larger than about $10^2$ M$_\odot$, if tidal disruption of stars is the principal source of the accreting matter on which it grows. In addition, the observational limits on the X-ray and gamma-ray continuum luminosity from the Galactic Center would be consistent (Figure 3) with an isothermal bremsstrahlung spectrum at T $\sim 1$ MeV and a luminosity of $\sim 2 \times 10^{38}$ erg/sec. Calculations[42] suggest that just such a temperature and luminosity is expected from optically thin disk accretion around a $\sim 10^2$ M$_\odot$ hole at a rate of $\sim 10^{-8}$ M$_\odot$/yr which would produce the hole in roughly the age of the Galaxy.

On the other hand, it may be possible to relax the size constraint on the photon-photon pair production volume, if we consider interactions of high energy beamed photons with one another (M. Burns, personal communication, 1982). In particular, in a beam of opening angle of the order of a degree, pair production by small angle interactions between beam photons could occur down to photon energies of tens of MeV and the fraction of the beam energy going into pairs could be of the order of $10^{-2}$. In such a case the beam photons and pairs might then be stopped in ambient gas with the bulk of the energy going into heating the gas by Compton and Coulomb collisions. This could then be reradiated isotropically as thermal emission with a luminosity of $\leqslant 10^{40}$ erg/sec, consistent with observational constraints on the $\leqslant 30,000$ K luminosity. Moreover since the radiation yield of $\leqslant 30$ MeV electrons and positrons is $\leqslant 3\%$, the hard X-ray and gamma-ray luminosity could also be less than the observational limit of $< 3 \times 10^{38}$ erg/sec.

The detailed energetics and geometries in both of these cases, however, are still under study.

## SUMMARY

In summary we find that the observations of the annihilation line emission from the direction of the Galactic Center, together with limits on the accompanying continuum emission, place important constraints on the nature of both the positron source and their annihilation region.

The observed line width of $< 2.5$ keV and time variations on a time scale of six months or less require that the annihilation region have a temperature of $\leqslant 50,000$ K, an ionization fraction of $\geqslant 10\%$, a probable density of $\sim 10^5$ H/cm$^3$ and a size of $< 10^{18}$ cm. Such conditions do in fact appear to exist in the peculiar warm clouds and other compact IR sources within the central parsec of the galaxy.

The limits on the accompanying continuum emission at energies $> m_e c^2$ appear to set the strongest constraints on the positron production process, requiring an exceedingly high efficiency such that roughly half of the total radiated energy $> m_e c^2$ goes into electron positron pairs. This constraint, however, must be qualified by assumptions that the positron production takes place on time scales comparable to that of the observed annihilation and in an optically thin, isotropically emitting region. Under these conditions only photon-photon pair production among roughly MeV photons can provide the required high efficiency. For isotropic photon-photon interactions the gamma ray continuum luminosity requires a highly compact source (d $< 5 \times 10^8$ cm). In this case pair production in intense bremsstrahlung emission from an $\sim$ MeV gas around an accreting black hole of $< 10^3$ M$_\odot$ appears to be a possible source. A larger source size may be possible, however, for photon-photon pair production by small angle interaction among high energy beamed photons. If the beam is then stopped in gas, the energy dissipated in heat may be radiated isotropically without violating the luminosity constraints. But if we relax the assumptions to consider the possibility that the accompanying high energy ($> m_e c^2$) emission may have escaped detection because the positron production occurred either in short intense bursts, in an optically thick medium, or in a highly collimated beam directed out of the line of sight, then there are still a number of alternative positron production processes. Clearly further study and observation is required.

## ACKNOWLEDGEMENTS

We wish to thank M. Burns, J.C. Higdon, D. Leiter, J.M. McKinley and S.E. Woosley for valuable discussions. The work of R.E.L. was supported by NASA grant NSG - 7541.

## REFERENCES

1. W.N. Johnson, F.R. Harnden and R.C. Haymes, Ap. J., *172* L1 (1972).
2. W.N. Johnson and R.C. Haymes, Ap. J., *184*, 103 (1973).
3. R.C. Haymes, et al., Ap. J., *201*, 593 (1975).
4. M. Leventhal, C.J. MacCallum and P.D. Stang, Ap. J., *225*, L11 (1978).
5. M. Leventhal et al., Ap. J., *240*, 338 (1980).
6. A.S. Jacobson, 10th Symp. Relativistic Astrophys., in press (1981).
7. G.R. Riegler, et al., Ap. J., *248*, L13 (1981).
8. R.W. Bussard, R. Ramaty and R.J. Drachman, Ap. J., *228*, 928 (1979).
9. C.J. Crannell et al., Ap. J., *210*, 582 (1976).
10. P.R. Shapiro and R.T. Moore, Ap. J., *207*, 460 (1976).
11. P.A. Sturrock and K.B. Baker, Ap. J., *234*, 612 (1979).
12. R. Ramaty and R.E. Lingenfelter, Nature, *278*, 127 (1979).
13. P.N. Okeke and M.J. Rees, A. Ap., *81*, 263 (1980).
14. J.L. Matteson, P.L. Nolan and L.E. Peterson, X-Ray Astronomy. (Pergamon Press, Oxford, 1979), p. 543.
15. G.R. Riegler et al., 17th Cosmic Ray Conf. Papers, *2*, 1 (1981).
16. I. Gatley and E.E. Becklin, Infrared Astronomy (Reidel, Dordrecht 1981), p. 281.
17. J.H. Lacy et al., Ap. J., *241*, 132 (1980).
18. E.E. Becklin et al., Ap. J., *219*, 121 (1978).
19. K.Y. Lo et al., Ap. J., *249*, 504 (1981).
20. M.G. Watson et al., Ap. J., *250*, 142 (1981).
21. T. Erber, Rev. Mod. Phys., *38*, 626 (1966).
22. P.A. Sturrock, Ap. J., *164*, 529 (1971).
23. A.F. Cheng and M.A. Ruderman, Ap. J., *216*, 865 (1977).
24. E.T. Scharlemann, J. Arons and W.M. Fawley, Ap. J., *222*, 297 (1978).
25. C.E. Fichtel et al., Ap. J., *198*, 163 (1975).
26. H.A. Mayer-Hasselwander, et al., Astron. Ap., in press (1982).
27. B.N. Swanenburg et al., Ap. J., *243*, L69 (1981).
28. D.D. Clayton, S.A. Colgate and G.J. Fishman, Ap. J., *155*, 75 (1969).
29. S.A. Colgate, Ap. Space Sci., *8*, 457 (1970).
30. F. Albernhe et al., Astron. Ap., *94*, 214 (1981).
31. R. Ramaty, B. Kozlovsky and R.E. Lingenfelter, Space Sci.Rev., *18*, 341 (1975).
32. R.E. Lingenfelter, J.C. Higdon and R. Ramaty, Gamma Ray Spectroscopy in Astrophysics (NASA, 1978), p. 252.

33. R. Ramaty, B. Kozlovsky and R.E. Lingenfelter, in preparation (1982).
34. R. Ramaty, B. Kozlovsky and R.E. Lingenfelter, Ap. J. Supp., *40*, 487 (1979).
35. R.D. Blandford, Active Galactic Nuclei (Cambridge Univ. Press, London, 1979), p. 241.
36. R.V.E. Lovelace, J. McAuslan and M. Burns, Particle Acceleration Mechanisms in Astrophysics (Am. Inst. Physics, New York, 1979), p. 399.
37. R. Ramaty and R.E. Lingenfelter, Phil. Trans. R. Soc. Lond., *A301*, 671 (1981).
38. L.F. Rodriguez and E.J. Chaisson, Ap. J., *228*, 734 (1979).
39. T.A. Weaver, Phys. Rev., 13A, *1563* (1976).
40. D. Lynden-Bell and M.J. Rees, M.N.R.A.S., *152*, 461 (1971).
41. L.M. Ozernoy, Large Scale Characteristics of the Galaxy (Reidel, Dordrecht, 1979), p. 395.
42. D.M. Eardley et al., Ap. J., *224*, 53 (1978).

# THE INTENSITY AND SPECTRUM OF GALACTIC CENTER $\beta^+$ ANNIHILATION PHOTONS AFTER COMPTON SCATTERING

D. J. Forrest
University of New Hampshire, Durham, NH 03824

## ABSTRACT

Measurements of the gamma ray spectrum from the Galactic Center region show two characteristics near 511 keV which are consistent with that expected from Compton scattering of a narrow line at 511 keV. The narrow line intensity at 511 keV has been observed to decrease by a factor of 2.8 over a six month period,[1] a factor which can be produced by an increase in optical thickness of $\sim 6$ g/cm$^2$. In addition to the narrow line, Compton scattered 511 keV photons will produce a spectrum with an $h\nu^{+1}$ energy dependence in the range 300-511 keV, a shape consistent with observations.[2,3] The calculated intensity of this scattered feature relative to the intensity of the unscattered narrow line for a spherical symmetric model is compared with $\beta^+$ annihilation via positronium and recent observations. Based on this comparison, it is concluded that a model involving Compton scattering cannot be excluded by present gamma ray observations.

## INTRODUCTION

Hard X- and gamma rays interactions with matter of typical astrophysical composition are completely dominated by Compton scattering over the broad energy range from a few keV to greater than 10 MeV. The physics of Compton scattering is well understood[4,5] and can be summarized by the following two equations for the energy and cross section of the scattered photon,

$$h\nu = h\nu_0 \left[1 + \frac{h\nu_0}{mc^2} (1 - \cos\theta)\right]^{-1} \quad (1)$$

$$\frac{d\sigma}{dh\nu} = \frac{\pi r_0^2 mc^2}{(h\nu_0)^2} \left\{ \frac{h\nu}{h\nu_0} + \frac{h\nu_0}{h\nu} - 2\left(\frac{mc^2}{h\nu} - \frac{mc^2}{h\nu_0}\right) + \left(\frac{mc^2}{h\nu} - \frac{mc^2}{h\nu_0}\right)^2 \right\} \quad (2)$$

where $r_0 = 2.8 \times 10^{-13}$ cm, and $\theta$ is the angle between the initial ($h\nu_0$) and the scattered ($h\nu$) photon. The important point shown by Equation 1 is that a scattered photon is always present after a Compton interaction, and hence, Compton scattering can be thought of as a wavelength shifting process which conserves photon number. Equation 2 gives the energy dependent cross section for these scattered photons and Figure 1 shows this dependence for $h\nu_0 = 511$ keV. We note that $d\sigma/dh\nu \sim h\nu^{+1}$ in the energy range 300-500 keV.

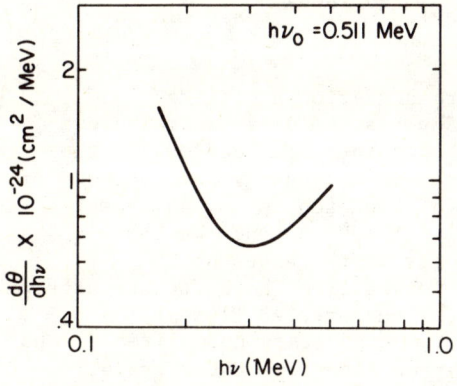

Fig. 1. Differential Compton scattering cross section for $h\nu_0 = 511$ keV.

There are three simple scattering geometries that can be considered. The first is a point source of radiation located above an infinite scattering plane, the second is a distributed source mixed uniformly within a spherically symmetric scattering medium (i.e., an "optically thick" source), and the third is a point source located within a finite but spherical symmetric scattering medium. Case 3 is considered in some detail in the next two sections.

## CALCULATIONS

The calculations assume spherical symmetry which is shown schematically in Figure 2 where L is the thickness (g/cm$^2$) of the scattering material. Furthermore, in recognition of the fact that experimental observations of the Galactic Center region are dominated at lower energies by a steeply falling continuum[2] (i.e., $E^{-2.5}$), we specifically calculate the scattered photon intensity only in the range 300-511 keV relative to the remaining narrow line intensity to compare with experimental data, as this is the energy range where the effect is most readily observed.

The integrated photon intensity in the range of 300-511 keV ($\Delta h\nu$) is given by,

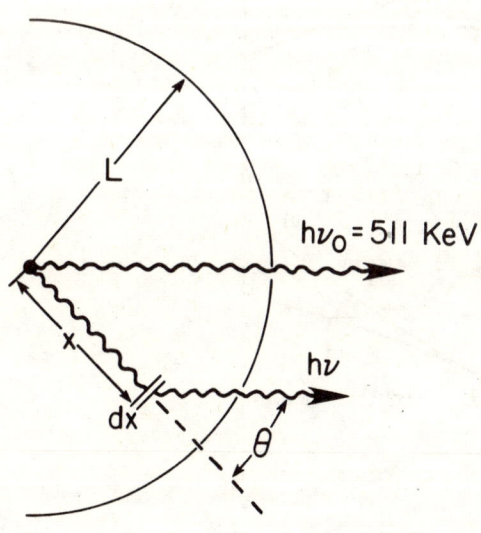

Fig. 2. Schematic picture of scattering geometry used in calculations.

$$I'(\Delta h\nu) \, \Delta h\nu \cong \int_0^L I_0 e^{-\sigma_s X} \, \sigma'(511,\Delta h\nu) \, e^{-\sigma_s(\Delta h\nu) \frac{L-X}{\cos \theta}} \, dX \, \Delta h\nu \quad (3)$$

where $\sigma_s$ is the total scattering cross section at 511 keV
(= $2.9 \times 10^{-25}$ cm$^2$/elect), $\sigma'(511,\Delta h\nu)$ is the differential cross
section for scattering a 511 keV photon into $\Delta h\nu$, $\sigma_s(\Delta h\nu)$ is the
cross section for scattering a photon in the energy band $\Delta h\nu$ out of
$\Delta h\nu$, and $\theta$ is the scattering angle of a photon in $\Delta h\nu$ as measured
from the trajectory of the original 511 keV photon. We have
numerically integrated Equation 3 over depth ranges up to
$L = 10$ g/cm$^2$. Figure 3a shows the intensity of both the unscattered
and scattered flux normalized to the unscattered flux at $L = 0$.
Figure 3b shows the ratio of these two quantities normalized to the
unscattered and hence observable narrow line flux after traveling
through a depth L. Also shown in Figure 3b is this same intensity

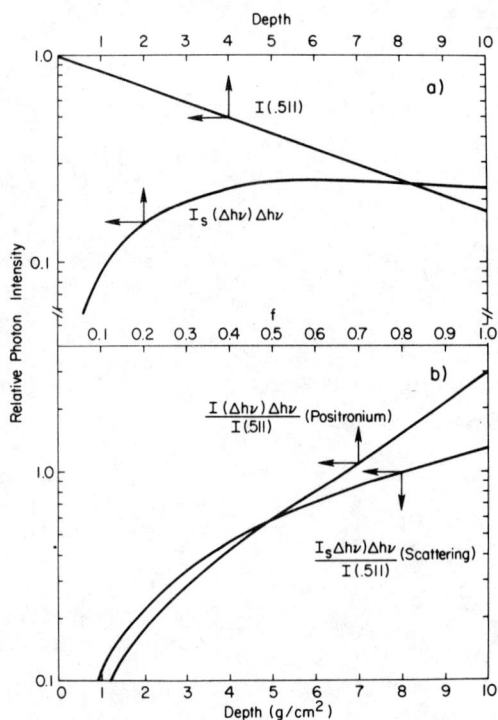

Fig. 3. Figure 3a is the photon intensity of the narrow line
[$I(h\nu_0)$] and the scattered continuum [$I(\Delta h\nu)\Delta h\nu$] for $h\nu_0 = 511$ keV
and $\Delta h\nu = 300$-511 keV. Figure 3b is the ratio of these intensities
both for scattering and the annihilation via positronium.

ratio calculated as a function of f, the fraction of positrons annihilating via positronium.[6] If f = 1, the annihilation produces two photons at 511 keV 25% of the time and 3 photon continuum with a $h\nu^{+1}$ spectral shape 75% of the time. This quantity is in general dependent on the local environment.[7,8] We have normalized this ratio to the observable narrow line intensity and assume no spectral changes due to scattering. Both processes produce similar $\sim h\nu^{+1}$ spectral shapes in the 300 to 511 keV range and the intensity ratios resulting from these two different processes are indistinguishable for $L \gtrsim 5$ g/cm$^2$ and $0.4 < f < 0.7$.

## INTERPRETATIONS AND COMPARISON WITH OBSERVATIONS

For a source not varying in time, observed intensity changes in the narrow 0.511 MeV line can be produced by changes in the scattering depth. For spherically symmetric geometries this can only be produced by continuous condensation of the scattering material, a process which may take considerable time. However, a more likely picture may be that represented by higher density scattering and absorbing regions in orbit around a central $\beta^+$ producing region. A specific example is represented by new star formation where the $\beta^+$ source region is the central core and the local high density region are "proto-planets". Equation 1 and Figure 2 shows that the paths followed by an observed unscattered and a scattered photon are markedly different. Hence, if a given observation was made while a density condensation was directly between the source and the observer, then the observed ratio could be much larger than that calculated for spherical symmetry because the narrow line is preferentially attenuated in this geometry. Spectral data to date has only been interpreted in the context of positronium formation. However, the spectral features resulting from allowed f's (i.e., $f \sim 0.8$)[2,3] could just as well be produced by Compton scattering in $\sim 5$ g/cm$^2$ of material.

We conclude that a physical model which includes scattering can explain all of the observed properties of the Galactic Center emission between 300 and 511 keV and that further theoretical and observable features of such a model should be investigated.

## REFERENCES

1. G. R. Riegler, J. C. Ling, W. A. Mahoney, W. A. Wheaton, J. B. Willett, A. S. Jacobson, and T. A. Prince, Astrophys. J. (Letters) <u>248</u>, L13 (1981).
2. M. Leventhal, C. J. MacCallum, and P. D. Stang, Astrophys. J. (Letters) <u>225</u>, L11 (1978).
3. B. M. Gardner, D. J. Forrest, P. P. Dunphy, and E. L. Chupp, Proceedings Workshop on the Galactic Center, California Institute of Technology, Pasadena, California, January 7-8, 1982.
4. R. D. Evans, The Atomic Nucleus (McGraw-Hill Book Company, New York, 1955).

5. A. T. Nelms, Graphs of the Compton Energy-Angle Relationship and the Klein-Nishina Fromula from 10 keV to 500 MeV, NBS-542, National Bureau of Standards, Washington, D.C., (1953).
6. M. Leventhal, Astrophys. J. (Letters) 183, L147 (1973).
7. R. W. Bussard, R. Ramaty, and R. J. Drachman, Astrophys. J. 228, 928 (1979).
8. C. J. Crannell, G. Joyce, and R. Ramaty, Astrophys. J. 210, 582 (1976).

# PHYSICS OF BLACK HOLES*

Kip S. Thorne[†]
California Institute of Technology

## ABSTRACT

The activity at the galactic center might be fuelled by energy release near a large black hole. In this talk I describe some relativistic effects which may be relevant to this process. I use Newtonian language so far as possible and illustrate the effects with "simple" analogies. Specifically, I describe the gravitational field near a black hole, Lens-Thirring and geodetic precession, electromagnetic energy extraction of the spin energy of a black hole and the structure of accretion tori around black holes.

---

*This paper is identical to "Black Holes and the Origin of Radio Sources", Kip S. Thorne and Roger D. Blandford, in "Extragalactic Radio Sources", Proceedings IAU Symposium #97, ed. D. S. Heeschen and C. M. Wade (Reidel, Dordrecht, Holland 1981).

[†]Supported in part by the National Science Foundation [AST79-22012]

# THE COMPACT SOURCE AT THE GALACTIC CENTER

Martin J. Rees
Institute of Astronomy, Madingley Rd., Cambridge, U.K.

## ABSTRACT

Various circumstantial arguments for (and against) the presence of a massive black hole ($\lesssim 5 \times 10^6$ $M_\odot$) are mentioned and assessed; in particular, the upper limits to the hole's mass based on considerations of stellar disruption and swallowing are shown to be very uncertain. If a massive black hole indeed lurks at the Galactic Center a compact radio source resembling the one actually observed would be a natural (indeed, almost inevitable) consequence of low-level accretion onto it. Related processes could account for an ionizing flux up to $\sim 10^{41}$ erg s$^{-1}$, and perhaps also for the observed $e^+ - e^-$ annihilation, but these latter aspects of the model involve physical uncertainties (particularly with regard to particle acceleration) which cannot be reliably quantified.

## INTRODUCTION

Are there any phenomena at the Galactic Center which resemble – in a muted form – those which we observe in external galaxies with active nuclei? This question is interesting if we are aiming to understand the data presented at this meeting. But it bears also on the general interpretation of galactic nuclei – whether, in particular, the Seyfert galaxies are just ordinary spirals undergoing a flaring phase.

The conjecture that there might be a massive black hole at the Galactic Center goes back at least 10 years[1]. The infra-red data from the Berkeley group[2,3] and recombination line observations[4] have in the last few years allowed us to rule out any mass above $\sim 5 \times 10^6$ $M_\odot$. There is no straight dynamical evidence _for_ a massive black hole – part at least of the central mass concentration could take some more conventional form (e.g. a star cluster), and in any case the observed gas may be flowing outward[5]. But the present data on the IR emission and on the unresolved radio source increasingly suggest that there is _some_ unique object right at the Center, even if much of the activity and structure outside the central parsec can be accounted for in conventional terms. Bailey[6] has shown that, if IRS 16 were simply a star cluster, then within a radius 0.1 pc the stars would be so closely packed that runaway dynamical evolution would take $\lesssim 10^9$ yrs. Unless we are observing at a special epoch in the history of our Galactic nucleus, this supports the idea that a massive central object may already have formed. A further suggestive argument is based on the conjecture that all normal galaxies pass intermittently through Seyfert phases. A massive black hole could then exist as a relic of earlier Seyfert-style outbursts, being "reactivated" whenever the fuelling rate is boosted[7]. If a massive black hole _is_ present in the central region of our galaxy, then dynamical friction would cause it to settle quickly into the true dynamical centre.

The compact radio source at the Galactic Center seems unique in our Galaxy — it is unlikely to be a pulsar, a radio star or a supernova. When we find a unique object in a unique location, it is not "ad hoc" to invoke a special explanation. In the second half of this paper I shall argue that low-level accretion onto a massive black hole provides a natural explanation of the compact radio source at the Galactic Center; moreover, the same model could actually be extended to account for the low-level activity seen in the nuclei of some nearby external galaxies. It could account also for IRS 16 and for the variable electron-positron annihilation line.

## STELLAR CAPTURE AND DISRUPTION: ARE THERE GOOD ARGUMENTS AGAINST A MASSIVE BLACK HOLE?

The fuelling rate required to power all the activity in the Galactic Center is very modest — $10^{-5}$ $M_\odot$ per year is more than enough. This fuel could consist of ordinary interstellar gas, augmented by debris from stellar disruptions or collisions (especially those involving giants) in the vicinity. Some authors, particularly Ozernoi and his collaborators[8-11], have argued against a black hole as massive as $10^6$ $M_\odot$ in our Galaxy on the grounds that the rate of capture of stars would be too high to be compatible with observations — the luminosity from the Galactic Center would exceed what is observed; moreover, the hole would gain mass by swallowing stars so that (it is claimed) it would be likely by now either to exceed the $5 \times 10^6$ $M_\odot$ upper limit, or still be $\leq 10^3$ $M_\odot$.

These estimates may be right; but I think the uncertainties involved are much greater than seems to have been appreciated in the published literature — so great in fact that one cannot, by this line of argument, marshall a convincing case against a massive black hole. The difficulties are of two kinds: reliably computing the stellar capture rate; and calculating the fate of the debris.

### The stellar disruption rate

To be disrupted, a star must pass closer to the hole than the tidal radius (or "Roche radius"). This radius obviously depends on the type of star, but in order of magnitude we have[12,13] $r_T = r_*(M_h/m_*)^{1/3}$. For solar-type stars

$$r_T \simeq 1.4 \times 10^{13} \, M_{h6}^{1/3} \text{ cm} \tag{1}$$

In this formula $M_{h6}$ denotes the hole's mass in units of $10^6$ $M_\odot$. The gravitational radius is $1.3 \times 10^{11}$ $M_{h6}$ cm, smaller by $\sim 100$ than $r_T$ for $M_{h6} \simeq 1$; fortunately, therefore, the tidal disruption occurs sufficiently far from the hole for Newtonian approximations to be adequate.

If a massive black hole were immersed in a star cluster, the initial rate of stellar disruption can be estimated from an "$n \sigma v$" argument[12] to be

$$\sim 10^{-3} \, M_{h6}^{4/3} \left( \frac{N_*}{10^6 \text{pc}^{-3}} \right) \left( \frac{V_*}{200 \text{ km s}^{-1}} \right)^{-1} \text{yr}^{-1}, \tag{2}$$

$N_*$ being the star density and $V_*$ being the velocity dispersion. Of course, the values of $N_*$ and $V_*$ are constrained by the virial theorem in terms of the radius of the assumed stellar system and the hole's mass. The relevant values of $N_*$ to be inserted in this formula is uncertain, particularly if some of the infrared luminosity is attributed to a massive central source rather than to stars. But there are further uncertainties. The presence of a "stellar cusp"(a population of stars in bound orbits around the hole) could in principle <u>increase</u> the capture rate[13], but this correction is small for $M \lesssim 5 \times 10^6 \ M_\odot$. On the other hand, (2) could be a serious <u>over</u>-estimate because of "loss cone" effects: once all the stars whose orbits pass close to the hole have been destroyed, further captures can proceed only insofar as dynamical interactions among the stars can repopulate the orbits. This process occurs on the relaxation time for the stellar system, which may be $\gtrsim 10^{10}$ yrs.

Fate of the debris

Even if we were to accept (2) as a valid estimate of the disruption rate, the observational consequences depend on the answers to three interlinked questions:

(i) What fraction of the debris goes down the hole, rather than being expelled?

(ii) What is the radiative efficiency for the accretion process? In other words, how many ergs of energy are radiated for each gram that is swallowed?

(iii) How long does it take to "digest" one star? In particular, how does the "flare duration" and decay timescale for such a process compare with the interval between one stellar disruption and the next?

According to Gurzadyan and Ozernoi[10], the answer to (i) is "100%". They argue that the debris forms a disc to which standard "thin accretion disc" theory can be applied; this implies an efficiency $\gtrsim 6\%$, and a timescale for the swallowing process sufficiently long that for $M_{h6} \gtrsim 1$ the luminosity due to one star never fades to an acceptable level before the next star is disrupted.

I should like to summarise an alternative view, according to which Ozernoi's answer to (i) could (but need not) be correct, but the luminous flare from each stellar remnant is briefer and less efficient then he estimates.

If a solar-type star, approching on an almost "parabolic" orbit with small impact parameter is disrupted by passage within a distance $r_T (\sim 10^2 r_g)$ of a $\sim 10^6 \ M_\odot$ black hole, the bits of debris move out along orbits whose mean binding energy (to the hole) is $GM_*/r_* \sim 10^{-5}$. This is because the energy needed to tear the star apart has come from the incoming orbital energy. These orbits are very eccentric, with typical major axes $\sim 10^5 \ r_g$. If the star approached somewhat closer than $r_T$, the disruption might be sufficiently violent (and the velocity spread within the debris sufficiently large) that some material might escape. But unless there is some explosive input of nuclear energy during the disruption (which would be expected only if a star passes <u>several times closer</u> than $r_T$(ref 14)) much of the material will remain <u>bound to the hole</u>. The "mean orbit" for the debris has a period of 50 $M_{h6}$ yrs.

For $M_{h6} \simeq 1$ this is very small compared even to the shortest estimates of the period between successive disruptions. The debris from each star is thus digested separately — in contrast to the "debris cloud"[11,12] expected in quasar models with higher $M_h$, where the disruptions are more frequent but the orbital periods of the debris ($\propto M_h$) are longer.

If the gaseous debris were completely cold, it would, after one or two orbital periods, form a highly elliptical disc, with a big spread in apobothric distance between the most and least lightly bound orbits, but where at peribothron all the orbits are squeezed into a range $(\Delta r/r) \simeq r_* \simeq 0.01\ r_T$. Gurzadyan and Ozernoi[10] show that applications of "$\alpha$-model" thin disc theory then yields a slow swallowing time.

We are, however, led towards a different picture (and maybe a more realistic one) if we note that at apobothron ($r \simeq 10^5\ r_g$) the characteristic speed of the debris, whose angular momentum corresponds only to that of a circular orbit at $r_T$ ($\lesssim 100 r_g$ for $M_{h6} \gtrsim 1$), is $\lesssim 30$ km s$^{-1}$. This is $\lesssim 3$ times the internal sound speed, on the assumption that the temperature is maintained by photoionization at $\gtrsim 10^4$ °K. Pressure gradients can then change the orbital angular momentum of elements of debris by factors of order unity. In consequence, when it falls again towards peribothron the debris will acquire <u>random</u> velocities of order $c(r_T/r_g)^{-\frac{1}{2}}$ - larger by a factor $(r_T/r_*) \simeq (M_h/m_*)^{1/3}$ than if each element had conserved the orbital angular momentum it had when it was part of the star. For $M_{h6} \simeq 1$ this gives an enhancement by $\sim 100$ in the random velocity; and therefore yields a $10^4$-fold increase in the viscous dissipation rate predicted by an "$\alpha$-model". Unless the effective $\alpha$ is $<< 1$, the debris will, after only a few periods, form an axisymmetric torus or "donut" (cf. ref 15), whose density maximum occurs in a ring with $r \simeq r_T$. The orbital period at $r_T$ is $\sim 1$ day; according to an $\alpha$-model the viscous dissipation timescale for a thick torus would be $\alpha$ times longer; even for $\alpha \simeq 10^{-2}$ this is less than 1 year. But we then come up against another limit. A mass $m_*$ in circular orbit at $r \simeq r_T$ cannot change its binding energy by a factor of order unity on a timescale shorter than

$$t_{Edd} = \left(\frac{4\pi Gmp}{\sigma_T c}\right)^{-1} \left(\frac{m_*}{M_h}\right)\left(\frac{r_g}{r_T}\right) \simeq 5 \times 10^8 \left(\frac{m_*}{M_h}\right)\left(\frac{r_g}{r_T}\right) \text{ yrs} \qquad (4)$$

For $M_h \simeq 10^6\ M_\odot$, this sets a minimum timescale of 5 yrs for a solar mass of material to deflate into a thin ring at $r \simeq r_T$, and $\sim 50$ years for it to be swallowed by the hole with $\gtrsim 10\%$ efficiency. If the viscous redistribution timescale, given by

$$t_{visc} = \alpha^{-1} t_{Kep}\left(\frac{h}{r}\right)^{-2}, \qquad (5)$$

is less than $t_{Edd}$, then a toroidal structure is set up, whose inner boundary moves in to the black hole. The material can then be swallowed from an orbit of small binding energy, implying a low efficiency.

For the above reasons, I would suggest that a disrupted star may be swallowed quickly and inefficiently — the only conspicuous luminosity being a flare of thermal optical and UV energy of $\sim 10^{44}$ erg s$^{-1}$, but only for a few months. Thus the present inactivity of our Galactic Center does not constrain the capture rate. What about the fraction of the mass which goes down the hole? I have already mentioned that some will escape after the initial disruption. Once the debris has developed into a torus supported by radiation pressure, then if the viscosity parameter $\alpha$ is $\ll 1$ and is a slowly-varying function of position, almost all material will eventually be swallowed (I am indebted to B. Paczynski for clarifying my ideas on this point); however, if the dissipation is sufficiently unsteady to give some fluid elements a higher entropy (i.e. a higher $(P/\rho)$), then those elements will be unbound. On general energetic grounds, we can see that only $\sim 10^{-4}$ of the mass, swallowed with high ($\gtrsim 10\%$) efficiency, could in principle release enough energy to expel all the remaining debris "to infinity".

[It may be worth noting parenthetically that there are advantages in the hypothesis that the mean output of the Galactic Center source — in ionizing radiation and mass efflux — may now be far below its mean value averaged over the last $10^4 - 10^5$ yrs.]

### TIDAL CAPTURE WITHOUT IRREVERSIBLE DISRUPTION: A POSSIBLE DIAGNOSTIC FOR A MASSIVE BLACK HOLE

A star which does not pass close enough to the hole to suffer disruption may nevertheless be tidally distorted; the associated dissipation brings the star into a tighter orbit, and further dissipation occurs at each peribothron passage until eventually the orbit is circularised at a radius a few times $r_T$ (refs 13, 16). As a result of this, the star would be in an orbit with period $\lesssim 1$ day (depending on type of star, but independent of $M_h$), but the orbital velocity would be

$$\sim c \left(\frac{r_T}{r_g}\right)^{-\frac{1}{2}} \simeq 0.1 \, M_{h6}^{1/3} \, c, \tag{6}$$

vastly higher than the speeds involved in ordinary binary systems. The gravitational radiation decay time from the tidal radius, for a solar-type star, is

$$\tau_{GR} \simeq 5 \times 10^6 \, M_{h6}^{-2/3} \text{ yrs}. \tag{7}$$

This is long compared with the interval between successive captures. Even though drag due to interaction with gas may degrade the orbit rather faster, any star which gets itself into a circular orbit near $r_T$ is likely to remain there until another star is captured on a similar orbit. This suggests that, if a massive black hole indeed lurks at the Galactic Center there is a 50% chance of there being a star in orbit around it. It would at least be worth looking for evidence of periodicity, due to the orbiting star itself or to its interaction with a torus/disc. Direct measurement of anomalously high orbital velocities would offer gratifyingly unambiguous evidence for a massive black hole. Such a stellar orbit would, furthermore, display interesting relativistic effects. For instance, if the central hole were spinning, and the star's orbital angular momentum were misaligned with respect to the hole's spin axis, then Lense-Thirring precession of the orbital plane would have a timescale of $(J/J_{max})^{-1}(r_g/c)(r_T/r_g)^3$, only of the order of years for the parameters appropriate to the Galactic Center.

But there is one uncertainty about this "tidal capture" mechanism. In acquiring a tightly-bound circular orbit of radius $\sim r_T$, the star must dispose of an amount of energy $\sim m_* c^2 (r_T/r_g)^{-1}$, larger by a factor $(M_h/m_*)^{2/3}$ than its entire original gravitational self-binding energy $\sim Gm_*^2/r_*$. The later stages of circularisation (occurring after the star has spun up to the circular velocity at peribothron) need to occur on a timescale longer than the star's thermal timescale - otherwise the star would be disrupted into a torus like that already discussed. This destruction need not be irreversible, however: a star could be "reborn" via gravitational instability in a dense gaseous ring after it had circularised. Alternatively, the tidal capture and circularisation process might be more gradual for a giant, where dissipation occurs in an extended envelope but a relatively dense core survives.

## LOW-LEVEL ACCRETION ONTO A MASSIVE HOLE: A MODEL FOR THE CENTRAL SOURCE

Unless we happen to be observing the hole very soon after it has disrupted a star, its fuelling rate is likely to be controlled by the infall of general interstellar gas. This rate is in itself uncertain, and probably variable, because of inhomogeneity of the physical conditions in the entire central regions; moreover there will be some "feedback" from the power output due to the hole's presence. But this fuelling rate will be "low", in the sense that it is far less than what would be needed to supply the "critical" or Eddington luminosity. Begelman, Blandford, Phinney and I have recently[17] discussed, in a different context, physical conditions around a black hole accreting at a very subcritical rate. I refer the reader to that paper for fuller details, and summarise the main points here, focussing on parameters that may be relevant to our Galactic Center.

The inflowing material is assumed to have sufficient angular momentum to prevent it from falling radially down the hole (the angular momentum orientation need not, however, for the present discussion, have any long-term constancy).

Shearing motions will maintain the magnetic field energy $B^2/8\pi$ at a significant fraction $\beta^{-1}$ of the pressure p. The Larmor radius is then, for the applications which interest us, about 10 orders of magnitude smaller than the scale of the flow. Ordinary 'molecular' viscosity will consequently be suppressed, and <u>magnetic</u> viscosity will probably govern the shear stress. Estimates suggest[18] that, in accretion discs, magnetic stresses may be $\alpha p$ with $\alpha \gtrsim 0.01$. Even though this estimate is uncertain, there is no reason why $\alpha$ should diminish as $\dot{M}$ falls: if $\alpha$ is fixed, the torque per unit mass is then independent of $\dot{M}$ (except insofar as the disc structure depends on $\dot{M}$)

Thus the inflow time in a disc of half-thickness h(r) around a hole of gravitational radius $r_g$

$$t_{inflow} = \alpha^{-1} \left(\frac{r}{r_g}\right)^{3/2} \left(\frac{h}{r}\right)^{-2} \left(\frac{r_g}{c}\right), \qquad (8)$$

does not depend explicitly on density. The <u>cooling time</u>, on the other hand, is inversely proportional to the density. When $\dot{M}$ is very low, the most likely flow pattern involves a "hot torus", with $h \simeq r$, supported by ion pressure with

$$kT_i \simeq m_p c^2 \left(\frac{r}{r_g}\right)^{-1}. \qquad (9)$$

When this is set up, (8) implies $t_{inflow} \simeq \alpha^{-1} t_{Kep}$ (cf. (5)). One then finds that for sufficiently low $\dot{M}$, $t_{inflow}$ becomes shorter than the timescale $t_{i \to e}$ for electron-ion coupling via Coulomb encounters. The condition for "weak coupling" is

$$\frac{\dot{M}}{\dot{M}_{crit}} < 50 \, \alpha^2, \qquad (10)$$

$\dot{M}_{crit}$ being the accretion rate yielding the Eddington luminosity for unit efficiency. When (10) is fulfilled, the ions can remain hot enough to satisfy (9) even if the electrons can cool.

<u>Some characteristic numbers</u>

The accretion-powered luminosity is approximately

$$L_{acc} \simeq 5 \times 10^{41} \dot{M}_{-5} \times (\text{efficiency}) \text{ erg s}^{-1}, \qquad (11)$$

where $\dot{M}_{-5}$ is the accretion rate in units of $10^{-5}$ solar masses per year. The efficiency for conversion of rest mass into luminosity is uncertain and may be rather low. (Just as in the radiation-supported tori discussed earlier, only a small fraction of rest mass can be disposed of radiatively during the timescale $t_{visc} \simeq t_{inflow}$ over which the material loses enough angular momentum to fall into the hole.) The energy release is concentrated in the part of the torus where the density peaks. The precise radius at which this happens depends on the angular momentum distribution in the torus (which is in turn determined by the uncertain viscosity), and on the value of $J/J_{max}$ for the hole. J is the hole's angular momentum, and $J_{max}$ the value corresponding to a "maximal Kerr" metric. A "typical" value would be $\sim 5$ times the gravitational radius $r_g$, i.e. a radius $r \simeq 7 \times 10^{''} M_{h6}$ cm. Characteristic values of relevant physical quantities are then:

$$n_e = 3 \times 10^{11} \alpha_{-2}^{-1} \dot{M}_{-5} M_{h6}^{-2} \left(\frac{r}{5r_g}\right)^{-3/2} \text{cm}^{-3}; \qquad (12)$$

$$B_{eq} = 4 \times 10^4 \, \alpha_{-2}^{-\frac{1}{2}} \, \dot{M}_{-5}^{\frac{1}{2}} \, M_{h6}^{-1} \left(\frac{r}{5r_g}\right)^{-5/4} \text{ G;} \qquad (13)$$

the electron-scattering optical depth would be roughly

$$\tau_{es} = 10^{-1} \, \alpha_{-2}^{-1} \, \dot{M}_{-5} \, M_{h6}^{-1} \left(\frac{r}{5r_g}\right)^{-\frac{1}{2}} \qquad (14)$$

In these expressions, $\alpha_{-2}$ denotes $10^{-2}\alpha$, where $\alpha$ is the viscosity parameter. The factors in brackets indicate how the various quantities would depend on r for $r > 5r_g$ (and in directions well away from the rotation axis).

### Energy balance for the electrons

Because, for low $\dot{M}$, $t_{i \to e} > t_{inflow}$, the energy of the ions cannot be drained away, and they remain at the virial temperature ($kT_i \simeq (r_g/r)$Gev) even if the electrons can cool. Note, however, that this aspect of the model would be invalidated if electron-ion coupling via collective effects were vastly more efficient.

Electrons at radii $r \lesssim 2000 \, r_g$ would be relativistic; and would cool via synchrotron and Compton emission. The synchrotron radiation emitted by electrons of Lorentz factor $\gamma_e$ comes out at frequencies $\propto \gamma^2 B$. The amount coming from radius r is limited by self-absorption to a luminosity proportional to $\gamma^7 r^2 (B(r))^3$. Compton scattering losses dominate synchrotron if $\gamma^2 \tau_{es} > 1$.

If there were no radiative losses at all, the electron Lorentz factor would increase adiabatically with decreasing r: for $r < 2000 \, r_g$ we would have

$$\gamma_{ad} \simeq \left(\frac{r}{2000 r_g}\right)^{-\frac{1}{2}} \qquad (15)$$

Note that even if the electrons cannot cool, their temperature everywhere within $r \simeq 2000 \, r_g$ falls below that of the ions: the ions, being non-relativistic, have an effective ratio of specific heats of 5/3 rather than 4/3, so if $n_e \propto r^{-3/2}$ their thermal energy varies as $r^{-1}$ rather than $r^{-\frac{1}{2}}$. For typical numbers appropriate to a Galactic Center model, (15) applies except for $r \lesssim 10 r_g$; at smaller values of r, $\gamma_e$ falls below $\gamma_{ad}$, levelling off and perhaps even decreasing as r decreases, because Compton losses are more efficient at small r where $\tau_{es}$ is higher (cf. (14)).

These considerations suggest a natural interpretation of the compact radio source. The properties of this source, as Dr Lo has described, are still poorly known, and there is only limited evidence on its spectrum. However, it seems to have (cf. ref 19) $L_{<8GHz} \simeq 10^{34}$ erg s$^{-1}$; the self-absorption limit at 8GHZ is $r \gtrsim 10^{13} B_c^{-\frac{1}{4}}$ cm.

The thermal electrons in the torus with $\gamma_e \lesssim 10$ at $r \gtrsim 10 \, r_g$ would naturally yield this radio emission (in more detailed estimates we must note that when optical depths are large the characteristic frequency of a single electron's emission may be at frequencies several times higher than $\nu_L \gamma_e^2$). If (15) applies, we calculate that the angular size of the source should vary with frequency as $\theta \propto \nu^{-4/9}$ and the spectrum should have the form $S(\nu) \propto \nu^{4/3}$. But these dependencies change if $\gamma_e$ differs from $\gamma_{ad}$. If $\gamma_e$ is independent of $r$, then $S(\nu) \propto \nu^{2/5}$. In general, we would expect synchrotron emission with a spectrum which rises quite steeply at low frequencies and then levels off. The frequency at which the spectral break occurs depends on $\dot{M}$, $\alpha$ and other uncertain parameters. The synchrotron spectrum would extend up at least into the infrared: for $\gamma_e \simeq 10$ and $B = 3 \times 10^4 G$ the optically-thin synchrotron spectrum peaks at $10^{12} Hz$, and for high optical depths to synchrotron self-absorption the peak moves up to $\gtrsim 10^{13} Hz$. Thus the most clear-cut observational consequences of low-level accretion onto a $\sim 10^6 M_\odot$ hole would be synchrotron-type emission in the radio and infrared, the latter coming from a source only a few times $r_g$ in size.

If this interpretation is basically correct, the Galactic Center is the only incoherent source where radio techniques are probing close to a relativistic object.

The linear polarization of the radio source would depend on the field configuration; since the relevant electrons have $\gamma_e \lesssim 10$, up to a few per cent circular polarization may be expected. However, this is very much an upper limit, because we would expect cancellations due to different parts of the source having opposite signs for $B_{||}$, and due also to self-absorption effects.

If high circular polarization were observed, this would suggest the more radical view that the radio emission was coherent. A "synchrotron maser" is well known to be impossible unless the plasma density or particle anisotropy are implausibly contrived. Coherent cyclotron emission, however, arises more readily; in fields $> 10^3$ G (cf. (13)) this would be in the GH band. The brightness temperature would then be $\gtrsim 10^{14}$ °K. Note that the more efficient cooling associated with coherent processes would reduce $\gamma_e$ below the values estimated previously (where only incoherent synchrotron and Compton losses were included) so it would not be inconsistent to invoke subrelativistic electrons. Such radiation would be subject to spontaneous (and induced) Compton scattering (cf. ref. 20); while this would be catastrophically effective if a quasar's radio luminosity came from dimensions of only a few times $r_g$, such need not be the case for the less extreme brightness temperatures appropriate to the Galactic Center.

## TOWARDS A MORE REALISTIC MODEL: SOME COMPLICATIONS

In discussing the state of the inflowing gas, we have assumed that at a given value of $r$ it is in a single "phase": the values of $T_i$ and $T_e$ can then be inferred by balancing radiative cooling against viscous and adiabatic heating. But realistically the state of the plasma may be very inhomogeneous.

(1) Some material may be in a cool phase at $\sim 10^4$ °K, the cool material constituting a thin disc or small dense cloudlets embedded in the hotter magnetised plasma.

(2) Some electrons may be accelerated to higher $\gamma$ by shocks, magnetic reconnection, etc. Such electrons would cool via synchrotron/Compton processes in a time $< r/c$. A "snapshot" of the torus would therefore reveal no more than a small fraction of the electrons to have $\gamma \gg \gamma_{ad}$. But these energetic electrons, radiating efficiently at high frequencies, may nevertheless make the main contribution to the luminosity of the torus. It is impossible to quantify these non-thermal acceleration processes. However, it would certainly be possible for the synchrotron/Compton spectrum to extend beyond the infra-red — indeed into the UV and even X-ray part of the spectrum. This contribution would increase the "efficiency" (cf. equation (11)).

If there were a smooth spectrum extending all the way from the far infra-red up to the ultraviolet, then it would be necessary to invoke extra extinction along the line of sight to IRS 16 as compared to the amount inferred for other sources, in order to achieve consistency with the detections or upper limits in the near infra-red. This extinction could be caused by a cool dusty outer disc (cf. ref. (1)).

## Can the $e^+ - e^-$ annihilation radiation be accounted for?

There are three mechanisms, none very compelling, whereby an ion-supported torus could generate electron-positron pairs.

(i) In the inner region, where $kT_i \gtrsim 50$ Mev, the protons (or, at least, those on the Maxwellian tail) may undergo collisions above the threshold ($\sim 290$ Mev) for $\pi^+$ production. Protons with these energies would of course be concurrently producing $\gamma$-rays from $\pi^0$ decay.

(ii) If a "non-thermal tail" of electrons are generated with $\gamma \gtrsim 10^3$, they can produce a synchrotron spectrum extending up to photons of $\sim 1$ kev. Compton scattering then yields $\sim 1$ Gev photons, which pair-produce via interaction with 1 kev photons before escaping (Blandford, this volume).

(iii) Electromagnetic processes near the hole may yield a cascade of $e^+ - e^-$ pairs (Lingenfelter and Ramaty, this volume).

All of these options (especially (iii)) have the weakness that the pairs are produced with high initial Lorentz factors. If their kinetic energy is degraded radiatively before annihilation occurs, then the predicted luminosity risks exceeding the acceptable upper limit from a central point source (even if the power is "hidden" in the far UV). The narrowness of the observed 0.511 Mev line sets limits on the Doppler broadening in the annihilation zone. Annihilation would have to occur in cool ($10^4 - 10^5$ °K) material[21] at $r \gtrsim 10^5$ $r_g$ (i.e. $r \gtrsim 10^{16}$ $M_{h6}$ cm), in a thin outer disc or in clouds embedded in hot plasma. If the pairs are produced near the central hole, they must be able to diffuse out to this distance in $\lesssim 1$ year. Furthermore, of course, they must not all annihilate quickly at smaller r. In the torus model this does not raise a problem, because at small r all the pairs would be maintained at relativistic energies by synchrotron absorption (unless coherent emission allows them to cool more efficiently); the cross section for annihilation would then be small

Relation to the nuclei of other galaxies

The distinctive pre-requisite for this type of quasi-stationary radio source is a low level of accretion (cf. eqn (10)). Qualitatively similarly behaviour may occur in other galactic nuclei. I would conjecture that those cases which do not display rapid variability may all be of this type, but that the variable nuclear components may involve "jets". The jets would be energised by extraction of the spin energy of the hole, coupled electromagnetically to the surrounding torus[17].

I have benefitted from discussions with a number of colleagues, and would particularly like to thank Mitch Begelman, Roger Blandford, Bob Brown, Juhan Frank and Sterl Phinney. I am grateful also to Guenter Riegler for the invitation to participate in the Galactic Center conference.

## REFERENCES

1. D. Lynden-Bell and M.J. Rees. MNRAS, 152, 461 (1971).
2. E.R. Woolman, T.R. Geballe, J.H. Lacy, & C.H. Townes. Astrophys.J. (Lett), 218, 103 (1977).
3. J.H. Lacy, F. Baas, C.H. Townes & T.R. Geballe. Astrophys.J.(Lett), 227, L17 (1979).
4. L.F. Rodriguez & E.J. Chaisson. Astrophys.J., 228, 734 (1979).
5. R. Brown. Astrophys.J. (submitted).
6. M.E. Bailey. MNRAS, 190, 217 (1980).
7. R.H. Sanders, Nature, 294, 427 (1981) and references cited therein.
8. L.M. Ozernoi in "Large-Scale Characteristics of the Galaxy", ed. W.B. Burton, p. 395 (Reidel, Holland, 1979).
9. V.I. Dokuchaev and L.M. Ozernoi. Sov.Astron.- A.J., 3, 209 (1977).
10. V.G. Gurzadyan & L.M. Ozernoi. Astron. Astrophys., 86, 315 (1980).
11. V.G. Gurzadyan & L.M. Ozernoi. Astron.Astrophys., 95, 39 (1981) and earlier references cited therein.
12. J.G. Hills. Nature, 254, 295 (1975).
13. J. Frank & M.J. Rees. MNRAS, 176, 633 (1976).
14. B. Carter & J. Luminet. Preprint (1981).
15. J. Frank. MNRAS, 187, 833 (1979).
16. A.C. Fabian, J.E. Pringle & M.J. Rees. MNRAS, 172, 15P (1975).
17. M.J. Rees, M.C. Begelman, R.D. Blandford & E.S. Phinney. Nature, 295, 17 (1982).
18. D.M. Eardley & A.P. Lightman. Astrophys.J., 200, 187 (1975).
19. S.P. Reynolds & C.F. McKee. Astrophys.J., 239, 893 (1980).
20. D.B. Wilson & M.J. Rees. MNRAS, 185, 297 (1978).
21. R.W. Bussard, R. Ramaty & R.J. Drachman. Astrophys.J., 228, 928 (1979).

# POSITRON PRODUCTION NEAR A $10^6$ $M_\odot$ BLACK HOLE

R. D. Blandford
Theoretical Astrophysics, Caltech, Pasadena, CA  91125

## ABSTRACT

The annihilation line coming from the direction of the galactic center may be produced by positrons originating from the magnetosphere of a $\sim 10^6$ $M_\odot$ black hole.

## INTRODUCTION

As discussed by Lingenfelter and Ramaty in these proceedings, one possible origin of the $\sim 10^{43}$ $s^{-1}$ positron source tentatively identified with the galactic center sources Sgr A, IRS16 and GCX is a compact region of very high radiation density. Under these conditions, γ-ray photons of energy $E_1 \gtrsim m_e c^2$ produce electron-positron pairs by interacting with photons of energy $E_2 \gtrsim m_e^2 c^4 E_1^{-1}$. The cross section for this process at threshold is $\sim 0.2\sigma_T \sim 10^{-25}$ $cm^2$ (Nikishov 1962). In this note, I wish to point out that the necessary efflux of positrons may emerge quite naturally from the magnetosphere of a magnetized $\sim 10^6$ $M_\odot$ black hole, as described at this meeting by Thorne and Rees.

Infrared observations of the galactic center suggest the presence of a central ionizing source of power $\sim 10^{40}$ erg $s^{-1}$ and effective temperature $\sim 30,000$ K (Gatley, these proceedings). This could be produced by an accretion disc of effective area $\sim 10^{26}$ $cm^2$ surrounding the black hole and being supplied with gas at a rate $\sim 10^{-5}$ $M_\odot$ $yr^{-1}$ (e.g. Shakura and Sunyaev, 1973). If this were so, the 10 eV photon density near the hole would be $\sim 10^{15}$ $cm^{-3}$. Now suppose that the horizon of the hole were threaded with a field of strength $10^4$ $B_4$ G. If the hole spun with an angular momentum J, an electric field of strength $\sim 3 \times 10^6$ $B_4 (J/J_{max})$ V $cm^{-1}$ would be induced in the magnetosphere surrounding the hole. Electrons and positrons could be produced by an electromagnetic cascade process (e.g. Blandford and Znajek 1977, Burns 1980). The basic breakdown mechanism is four stage.

(i) electron (or positron) is accelerated in electrostatic field, induced from the magnetic field by the rotation of the hole, to an energy $E_\gamma$

(ii) energetic electron or positron inverse Compton scatters a soft photon of energy $E_{uv} \sim m_e^2 c^4 E_\gamma^{-1}$ to make a gamma ray of energy $\sim E_\gamma$

(iii) gamma ray interacts with a second soft photon to produce an electron and a positron each of energy $\sim .5 E_\gamma$

(iv) electron and positron cool rapidly by the synchrotron process from an energy $\sim .5 E_\gamma$ to $\sim m_e c^2$.

The cycle then repeats until enough pairs have been produced to short out the electric field component parallel to the magnetic field.

Necessary conditions for this breakdown to occur are that the magnetosphere be optically thick in the soft photons and that the electric field be strong enough to accelerate the electrons and positrons to an energy $\sim E_\gamma$. In our example, the optical depth in the UV photons is $\sim 30$ and $E_\gamma \sim 10^5$ MeV. The maximum energy to which we could accelerate electrons or positrons can be estimated by balancing electrostatic acceleration with radiation reaction. This gives $E_\gamma \lesssim 3 \times 10^6 B_4^{\frac{1}{2}} (J/J_{max})^{\frac{1}{2}}$ MeV. Therefore both of these conditions are satisfied. The total electromagnetic power extracted from a $\sim 10^6 M_\odot$ black hole is $\sim 10^{40} (J/J_{max})^2 B_4^2$ erg s$^{-1}$. It is not clear what fraction of this power emerges from the magnetosphere in the form of relativistic particles. This depends upon the effective resistivity of the plasma within the magnetosphere which can only be guessed. However, suppose that a fraction f of this power goes into pairs of energy $E_\gamma \sim 10^5$ MeV and that $B_4 \sim 1$. These primary particles will be created with large pitch angles and will rapidly cool, radiating a synchrotron power $\sim 10^{40}$ f erg s$^{-1}$ in the form of $\sim 1$ MeV gamma rays. These gamma rays could produce a far larger flux of secondary positrons than is associated with the primary particles. The $\sim 1$ MeV gamma ray density in the magnetosphere is $\sim 10^{13}$ f cm$^{-3}$ which implies that a fraction $\sim 0.1$ f of them will be converted into electron positron pairs. The total positron efflux from the magnetosphere would then be $\sim 10^{45}$ f$^2$ s$^{-1}$ and so we could account for the observed strength of the annihilation line if f $\sim 0.1$.

A spinning, magnetized hole can therefore produce two outflowing beams of mildly relativistic pairs and MeV γ-rays. Positron annihilation will not occur until this beam has been stopped, which observations mandate to be at a distance $> 3 \times 10^5$ Schwarzschild radii $\sim 10^{17}$ cm. The observed variability imposes an upper bound on this distance $\sim 3 \times 10^{17}$ cm. As described by Dr. Ramaty, this annihilation must occur in a partially ionized and comparatively dense gas cloud. Could this be IRS16, displaced by $\sim 1"$ from the compact radio source? The MeV γ-rays produced by the magnetosphere would mostly be beamed parallel to the field and so the spin axis of the hole should not be directly observable at Earth. The electrons and positrons could possibly supply the compact radio source.

Our ignorance of the actual physical conditions within a hypothetical black hole magnetosphere is even deeper than is the case for pulsar magnetospheres where we have comparatively good diagnostics. The foregoing estimates must therefore be regarded as highly uncertain. However, I hope that I have demonstrated that the galactic center positrons could be produced from the environs of a $\sim 10^6 M_\odot$ black hole in a manner which is electrodynamically and radiatively self consistent.

## ACKNOWLEDGEMENTS

I am indebted to Drs. Lingenfelter, Paczynski, Ramaty and Rees for constructive criticism of an earlier version of these ideas. I gratefully acknowledge financial support by the National Science Foundation under grant AST80-11752 and the Alfred P. Sloan Foundation.

## REFERENCES

1. R. D. Blandford and R. L. Znajek, Mon. Not. R. astr. Soc. 179, 433 (1977).
2. M. Burns, unpublished thesis, Cornell University (1980).
3. A. I. Nikishov, Sov. Phys. JETP 14, 393 (1962).
4. N. I. Shakura and R. A. Sunyaev, Astr. Astrophys. 24, 337 (1973).

## Chapter VII Dynamics of the Galactic Center

### GAS MOTIONS IN THE CENTRAL REGION AND THEIR INTERPRETATION

J.H. Oort
Sterrewacht te Leiden, Leiden, The Netherlands

#### ABSTRACT

Three regimes may be distinguished: that within $\sim 1$ pc from the center containing fast-moving compact HII regions, observed principally in the NeII line at 12.8 µm, the region within $\sim 300$ pc with its very high density of molecular clouds, and the region between roughly 1 and 3 kpc containing extended features the motions of which deviate widely from circular motions. The compact HII clumps must be short-lived. No entirely satisfactory explanation of their origin has yet been found. The non-circular motions and the tilted distribution of the large HI features have alternatively been ascribed to eruptive activity from the nucleus and to large-scale streamings in a non-axisymmetric potential field. Recent studies of orbits in three-axial ellipsoids have shown the existence of "anomalous" orbits in tilted planes which may explain at least part of the observed features. It is doubtful whether the motions of the molecular clouds within 300 pc can be interpreted in a similar manner. Strong asymmetries and the existence of a ring-like expanding feature point rather to expulsion from the nucleus.

----

There are three classes of phenomena that require explanation:
  I  The large radial motions of major portions of the gas
  II The tilted distribution of the gas
  III Major asymmetries
It will be expedient to distinguish between three regions:
(a) The "nucleus", within $\sim 1$ pc from the center,
(b) The domain of the dense molecular clouds, within $\sim 300$ pc,
(c) The region between roughly 1 and 3 kpc from the center.
  The "nucleus" contains a very strong concentration of compact HII clumps which have been observed in great detail in the NeII fine structure line at 12.8 µm. Lacy has reported on these. It contains also a steep concentration of bulge stars which can be observed in the nearer infrared, between $\sim 2$ and 3 µm, as well as the ultra-compact radio source discussed by Lo and Rees.
  The region within a few hundred parsecs radius is unique because of the extremely high density of gas, which is mostly in molecular form. Part of the gas seems to move in roughly circular orbits, but other parts deviate widely from such a pattern. At least one ring-like feature is observed; this seems to move away from the center at velocities around 150 km s$^{-1}$. The distribution of the molecular clouds (observed in detail in the CO line at 2.6 mm, by Bania, Burton and Liszt, and others) is very asymmetrical; for instance, the density between 0° and about +1°5 longitude is very much higher than at the corresponding negative longitudes (cf. Fig. 1).

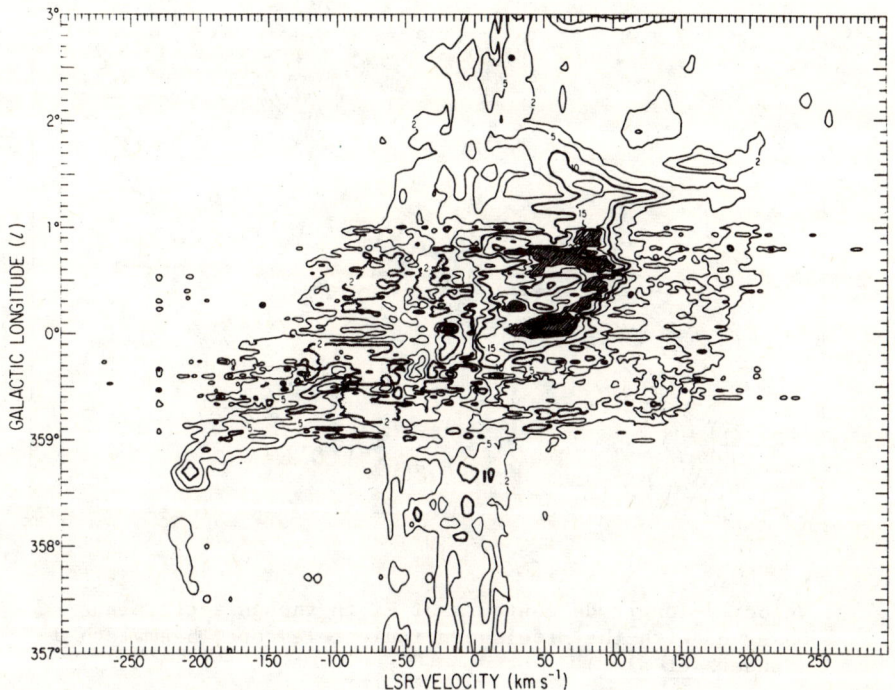

Fig. 1. CO distribution in velocity and longitude at b = 0° (from observations by Bania[1] and by Liszt et al[2]).

The distribution and motion of neutral hydrogen in longitude and velocity between -3° and +3° longitude, at zero latitude (Fig.2) shows that at $\ell = 0°$ HI occurs in absorption against Sgr A up to almost -200 km s$^{-1}$, thus moving away from the center over a wide velocity range; the gas <u>behind</u> the center, seen in emission, moves also away from the center, at velocities ranging similarly up to almost 200 km s$^{-1}$. When we combine these data with observations at other longitudes we see that this seemingly "expanding" gas is part of large structures (cf. Oort[4], Table 2 and Figure 11) far outside the region of the massive molecular clouds, and extending to several kpc from the center (Fig. 3). The much discussed 3-kpc arm is the most massive of these features.

The gas within about 1 kpc lies in a strongly tilted layer, as is illustrated in Fig. 4, which is reproduced from an investigation by Liszt & Burton[5].

Ever since the discovery of the 3-kpc arm in 1957 there has been a controversy about the interpretation of its outward motion: whether this is due to expulsion from the nucleus or to a deviation from axial symmetry in the gravitational field. The present tendency is to favour the latter interpretation. But there is no proof as yet.

Several investigators have attempted to represent the gas motions by general flow patterns. The most complete is the "barlike

182

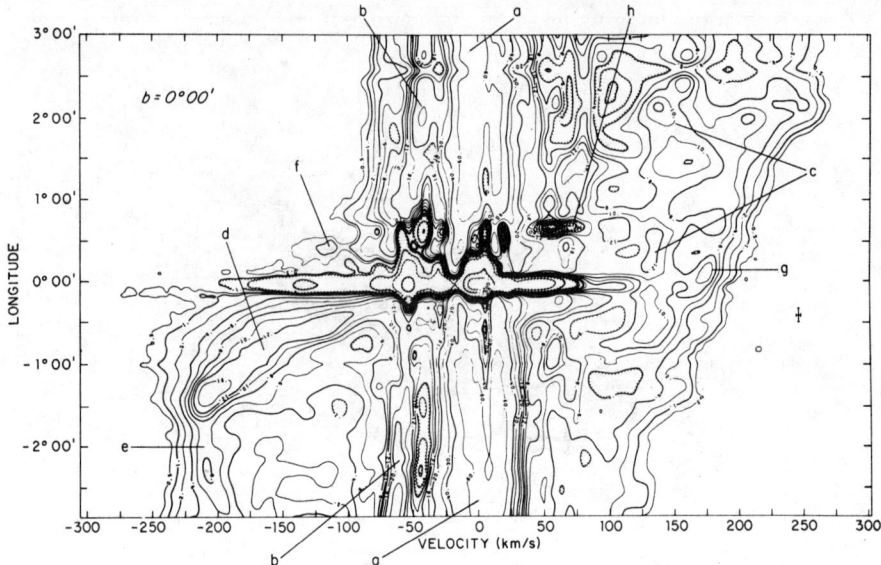

Fig. 2. Velocity-longitude contours of HI in the galactic plane observed with the Effelsberg radio telescope; beamwidth 9' (Sanders et al [3]).

model" designed by Liszt & Burton[5] to represent their very extensive and precise observations. In this model the gas is supposed to lie in a disk tilted 24° relative to the galactic equator; in this disk it moves in central ellipses with major axes in the same direction and axial ratios ranging between 1.6 for the innermost to 3.1 for the outermost orbits; the latter extend to R = 1.9 kpc. Burton and Liszt succeeded in finding model parameters by which most of the features that had been observed in HI - in particular the large outward radial motions in Fig. 3 - as well as some of the features of the molecular clouds could be well represented. The authors made no attempt to find a mass model which could produce closed orbits with the required characteristics.

Recently Lake and Norman[6] have computed a potential in which such orbits could occur. It remains to be seen in how far their artifical potential can be reconciled with the density distribution derived from infrared observations or from the light-distribution in M31.

An extremely interesting new development, started by Schwarzschild, promises to provide at least part of the physical basis that has so far been lacking. It has shown that both the tilt and retrograde motions may be quite natural phenomena.

Suppose the mass distribution of the central bulge is tri-axial and symmetrical with respect to the three principal planes. In the

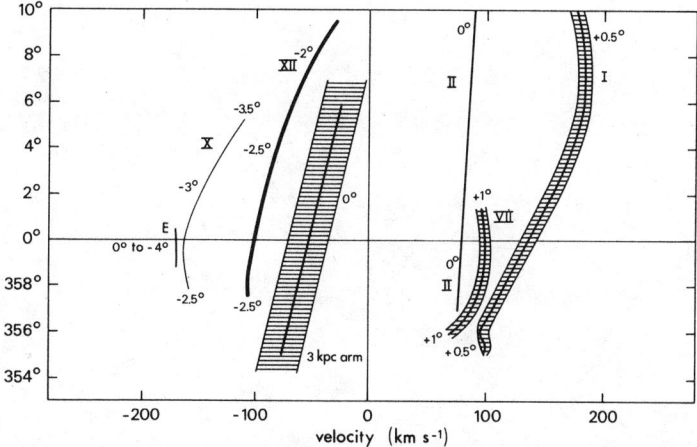

Fig. 3. Principal "expanding" features. Small figures indicate average galactic latitudes. Width of the shading or thickness of the curve is roughly proportional to the HI mass per unit interval of longitude (Oort[4], Figure 11).

case of gas motions we are primarily interested in stable closed orbits that do not self-intersect. Analyses of orbits in a non-rotating tri-axial system have shown that closed stable orbits exist in planes perpendicular to the short and long axes, and that these orbits have nearly elliptical shapes (Schwarzschild[7], Heiligman & Schwarzschild[8]). Gas disks can therefore exist perpendicular to the short axis and perpendicular to the long axis.

Now allow the figure to rotate slowly around its short axis. A gas disk perpendicular to the short axis would not be greatly affected by this rotation. If, however, the gas rotated in a plane perpendicular to the long axis the orbits would be changed in an important way. They become tilted with respect to the plane perpendicular to the long axis, the tilt increasing with the orbital radius R until for large R they approach the equatorial plane (the plane perpendicular to the short axis). The orbits of this family, which Schwarzschild has named anomalous orbits, have been shown to be periodic and closed. They exist over a considerable range of R. Numerical computations have been made by Merritt for a specific potential with axial ratios x:y:z of 2:1.25:1, and for a specific angular speed of the figure rotation. The anomalous orbits are nearly planar and fairly circular. The sense of their tilt is such that the orbital motion is retrograde relative to the figure rotation when viewed down the rotation axis. They could therefore offer a natural explanation for the observed tilt of the gas layer as well as for the "forbidden" (i.e. directed opposite to the general galactic rotation) velocities.

Van Albada, Kotanyi and Schwarzschild[9] have shown that the anomalous closed orbits in tri-axial density distributions give a natural explanation of the dust lanes perpendicular to the long axes

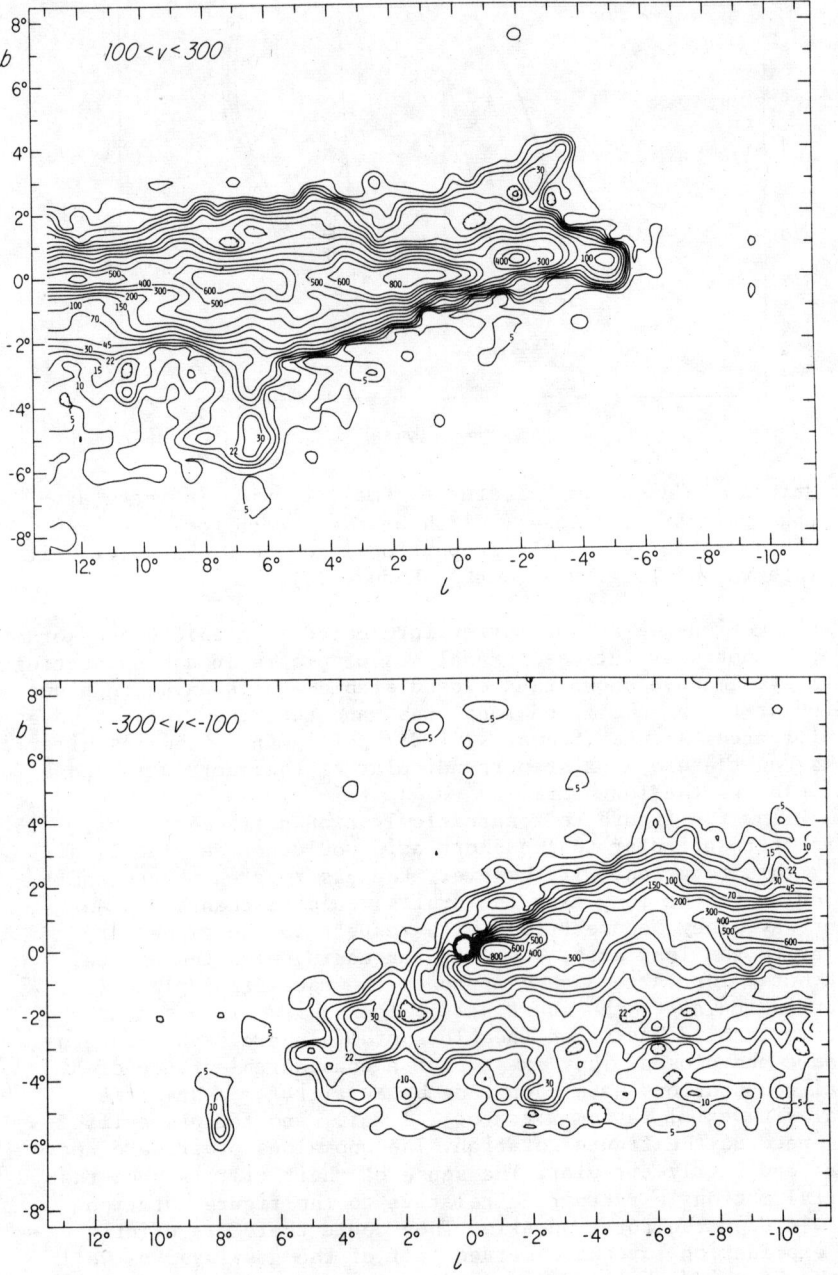

Fig. 4. Contours of observed antenna temperature integrated over the velocity range 100-300 km s$^{-1}$ (Liszt & Burton[5]).

that have been observed in some elliptical parent galaxies of radio sources, such as NGC 5128 in Centaurus A. They have also shown how the figure rotations can give rise to the symmetrical bending of the radio contours at large distances which have a long time been such a puzzling phenomenon.

T. de Zeeuw and W.A. Mulder in Leiden are now engaged in computing realistic models which can reproduce the complicated motions observed in the central region of our Galaxy. I am obliged to them for the information given above.

It should be pointed out that the nearly circular retrograde orbits considered would not explain the large motions directed away from the center observed in the line profiles of Sagittarius A. To fit these into the closed-orbit picture would require highly excentric prograde orbits, such as those of Burton and Liszt's model.

It is tempting to consider a tri-axial density distribution and a slow rotation of its figure as an explanation for some of the most striking anomalies in the central region: the tilted gas distribution and the retrograde motions.

But the reality is undoubtedly much more complicated. On the one hand there are the largest features like the 3-kpc arm, the non-circular motion of which may most easily be explained as due to a relatively slight bar-like disturbance such as considered by Sanders[10].

On the other hand there are the striking anomalies seen in the "molecular" region within 300 pc, which do not fit at all in the schematic models considered. Examples are: (a) the so-called "+40 km s$^{-1}$ cloud" which produces such a wide and deep absorption band in OH and $H_2CO$ in Sgr A (Fig. 5), and has a structure which extends almost exclusively to the side of positive longitudes (Fig. 6);

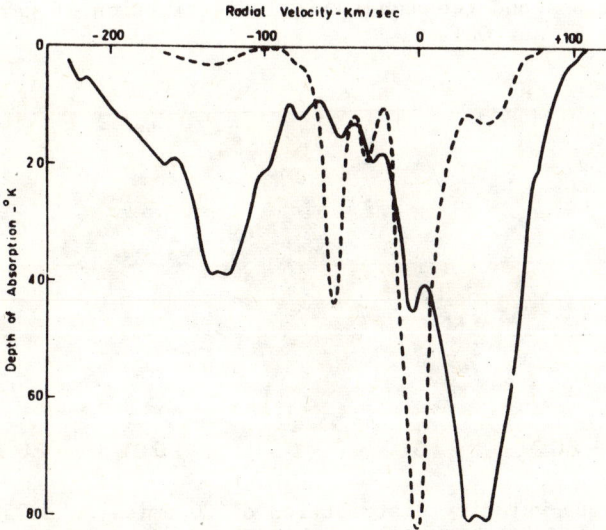

Fig. 5. Absorption profile of Sgr A. Full curve: OH, showing absorption by the +40 km s$^{-1}$ cloud and by the expanding ring at -130 km s$^{-1}$; dashed curve: HI (Bolton et al[11]).

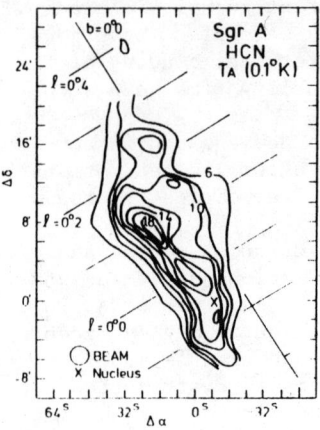

Fig. 6. Contours of maximum antenna temperature of HCN emission from the +40 km s$^{-1}$ cloud (Fukui et al [12]).

(b) the extremely asymmetrical distribution of the dense molecular gas within 2° from the center and its large deviation from circular motion; (c) the ring of molecular gas which between R ∼ 100 and ∼ 200 pc moves away from the center at ∼ 150 km s$^{-1}$ velocity (cf. Fig. 1). All these features are strongly tilted to the galactic plane, as may be illustrated by Fig. 7, copied from Liszt & Burton[13]. Phenomena such as (a) and (b) must be very short-lived recent formations. The circulation around the center would wipe them out in a few million years. They give the impression of being due to expulsive phenomena, in which quite large masses must have been involved.

In view of the features observed in spiral galaxies of the Seyfert type the occurrence of expulsion is nothing unexpected. In some spirals expulsions occur that are so powerful that they affect the structure over the entire disk; the remarkable anomalous arms in the bright spiral NGC 4258 present a striking example (Van der Kruit et al[14]). In summary we conclude that at least two different mechanisms: a central bar-like deviation from axial aymmetry and explosive activity of large scale must have been involved in producing the anomalous motions and the asymmetrical distribution of gas in the cental region of our Galaxy.

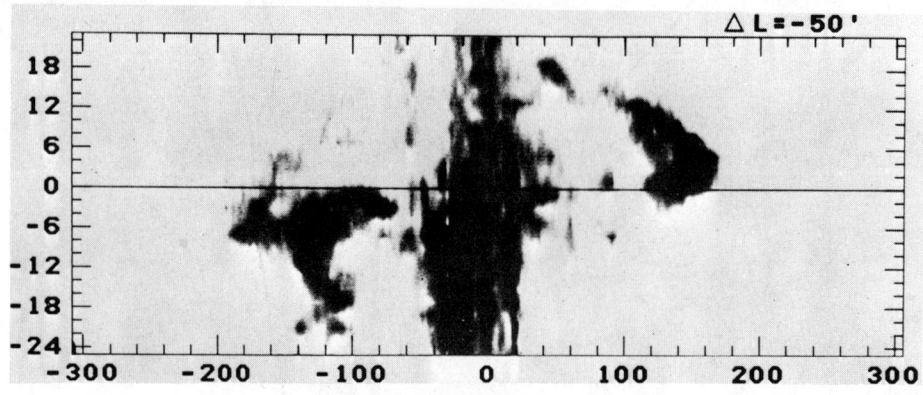

Fig. 7. Latitude-velocity distribution of CO emission at a longitude of -50' from the galactic center, showing tilt of the gas layer (Liszt & Burton[13]).

One important problem must still be discussed, viz., the determination of the mass density. The manner in which this varies with R, as well as the projected axial ratio have generally been determined from measurements of the radiation between 2 and 3 μm. In order to convert the 2.2 μm emission into mass density we need to know the extinction in front of the center and the M/L ratio at the wavelength concerned. The former can now be fairly well determined, but the latter remains uncertain. A calibration by means of dynamical data is therefore essential. For R > 40 pc this has generally been done by means of the rotational velocity of the nuclear disk. This is made somewhat uncertain by the evident deviations from circular motion in at least part of the disk, such as have been discussed in the foregoing. An entirely independent calibration has been made from the velocity dispersion and density distribution of bulge objects near the center, such as OH masers and planetary nebulae (cf. Isaacman[15]). These indicate that former estimates of the mass within R = 1 kpc might have to be reduced by about 25%; but due to the small number of available objects the calibration has no great accuracy.

As regards the mass of the small nucleus the most pertinent observations are those concerning the motions and distribution of the 14 compact HII regions within R = 1.5 pc that have been found at wavelengths around 10 μm. Their velocities have been determined from the fine structure line of NeII at 12.8 μm. The velocity dispersion is $126 \pm 20$ km s$^{-1}$. It may be higher in the inner 0.3 pc; 6 clouds within R = 0.3 pc give $\sigma \sim 190$ km s$^{-1}$. The space density $\nu$ is found to vary steeply: $\nu \propto R^{-\beta}$, with $\beta = 3.5 \pm 0.5$. On the assumption that motions and density distribution are governed by gravitation the mass within R = 0.5 pc is estimated as $\sim 5 \times 10^6$ $M_\odot$ (cf. Lacy et al[16]). This is roughly 3 times the mass estimated from the 2.2 μm radiation as calibrated by means of the disk rotation, and might therefore be taken as indicating the existence of a dark mass of several million solar masses. But the result is extremely uncertain; principally because we do not know how the fast-moving compact HII regions have been formed. They might be so young that there has not been time to reach a relaxed distribution.

Thus it is still unknown whether the galactic nucleus contains an ultra-compact mass of the same nature as those which are presumed to form the central engines of radio galaxies.

## REFERENCES

1. T.M. Bania, Ap.J. 216, 381 (1977).
2. H.S. Liszt, W.B. Burton, R.H. Sanders, N.Z. Scoville, Ap.J. 213, 38 (1977).
3. R.H. Sanders, G.T. Wrixon, U. Mebold, Astron.Astrophys. 61, 329 (1977).
4. J.H. Oort, Ann.Rev.Astron.Astrophys. 15, 295 (1977).
5. H.S. Liszt, W.B. Burton, Ap.J. 236, 779 (1980).
6. G.R. Lake, C.A. Norman, this volume (1982).
7. M. Schwarzschild, Ap.J. 232, 236 (1979).
8. G. Heiligman, M. Schwarzschild, Ap.J. 233, 872 (1979).
9. T.S. van Albada, C.G. Kotanyi, M. Schwarzschild, Monthly Not.Roy. Astr. Soc. 198, 303 (1982).

10. R.H. Sanders, in The Large Scale Characteristics of the Galaxy, (I.A.U. Symposium No. 84, Ed.W.B.Burton, 1979), p.383.
11. J.G. Bolton, F.F. Gardner, R.X. McGee, B.J. Robinson, Nature $\underline{204}$, 30 (1964).
12. Y. Fukui, T. Iguchi, N. Kaifu, Y. Chicada, M. Morimoto, K. Nagane, K. Miyazawa, T. Miyaji, Publ.Astr.Soc. Japan $\underline{29}$, 643 (1977).
13. H.S. Liszt, W.B. Burton, Ap.J. $\underline{226}$, 790 (Fig. 11b) (1978).
14. P.C. van der Kruit, J.H. Oort, D.S. Mathewson, Astron.Astrophys. $\underline{21}$, 169 (1972).
15. R. Isaacman, Astron.Astrophys. $\underline{95}$, 46 (1981).
16. J.H. Lacy, C.H. Townes, D.J. Hollenbach, preprint (1982).

# TRIAXIALITY AND THE GALACTIC CENTER

George R. Lake
Bell Laboratories, Murray Hill, New Jersey 07974

Colin Norman
Huygen's Laboratorium, Leiden, The Netherlands

## ABSTRACT

We investigate the properties of triaxial galaxies (those without axial symmetry) relevant to observations and inferred physical processes at the Galactic center. The features we find are: (1) velocity fields which mimic that of a supermassive object at the center and appear to decline more steeply with radius than Keplerian at larger radii, (2) the existence of stable elliptical orbits which are tilted with respect to all the symmetry planes, a possible explanation for the tilted disk observed at the galactic center, (3) the possibility of enhanced fueling due to the disappearance of all angular momentum integrals and hence the absence of any loss-cone.

## INTRODUCTION

There is a wide variety of reasons to consider the dynamics of triaxial galaxies. Recent observations and theoretical work suggest that many, if not most, elliptical galaxies are triaxial. Elliptical galaxies are not rotationally flattened[1] and their isophotal profiles are often twisted.[2] Residuals from the Faber-Jackson relation (luminosity proportional to the velocity dispersion to the fourth power) examined together with observed axial ratios indicate that elliptical galaxies are nearly prolate.[3] There are many elliptical galaxies with dust lanes in configurations which are not stable in axisymmetric systems.[4] N-body simulations of collapse have produced a number of final state figures which are triaxial. Even those still skeptical will have to grant that barred spirals (a category which seems to include the Galaxy) are certainly not axisymmetric.

Recently we have considered both stellar and gaseous dynamics of triaxial galaxies.[4] In this brief report, we present the features relevant to the Galactic center. The next section details the phenomena described in the abstract.

## RESULTS

a) Velocity Fields

We have explicitly calculated closed elliptical orbits in the plane of the model potential proposed by Schwarzschild.[5] These are shown in Fig. 1. Here X is the longest axis and Y is the intermediate axis (the orbits are all calculated in the Z = 0 plane). Note that the orbits counteralign to the figure, i.e., their long

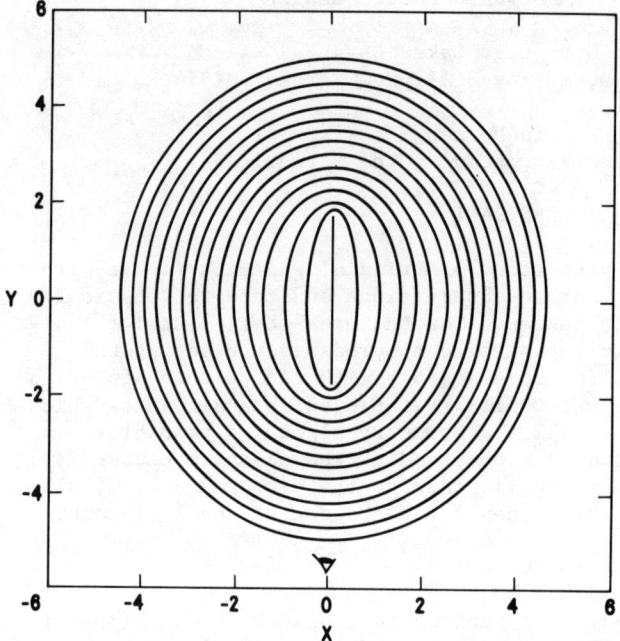

Fig. 1. Closed elliptical orbits in the model proposed by Schwarzschild.

axis is the Y axis. At low amplitude the spring constant in the Y direction is stiffer than that in X. At larger amplitudes the non-linearity of the potential allows for locking of the X and Y frequencies. At this point the Y-axial orbit becomes unstable and the elliptical orbits emerge at the bifurcation point.

Examining Fig. 1, one may see that the changing eccentricity of

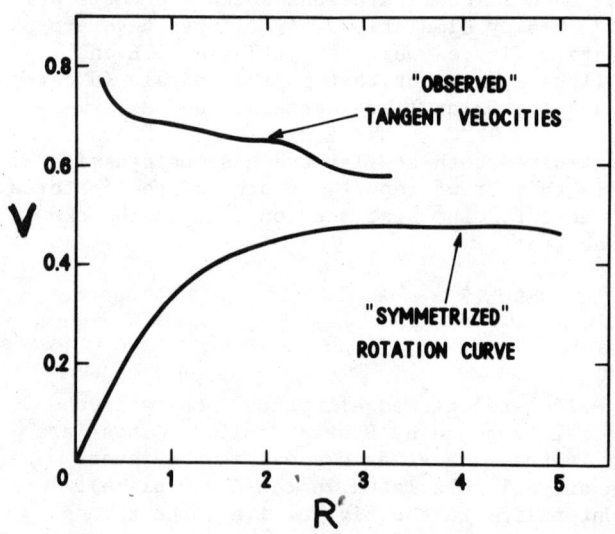

Fig. 2. Rotation curves for Schwarzschild model. Top curve is derived by sighting through orbits shown in Fig. 1. Bottom curve is given by $\sqrt{(F_x+F_y) \cdot r}$ where $F_x, F_y$ are the forces on the X and Y axes at a distance r from the center.

the orbits can produce interesting effects in the observed rotation curve. Fig. 2 shows sample rotation curves from a "computer telescope." Also shown is the rotation curve of the symmetrized potential. Note the rise in the center and unusually rapid declines which are possible. Such features in the past have been assumed to be due to supermassive objects[6] and disks with holes producing a ring which tugs outward.[7] This latter explanation is probably not possible in any case. If the rotation velocity really declined faster than Keplerian, the epicyclic frequency would be imaginary; i.e., there would be a kinematic Rayleigh instability. We should instead take such features to be sure indicators of noncircular motions.

b) Tilted Disks (and a word on precession of jets)

In the innermost 600 parsecs of the Galaxy, Liszt and Burton[8] report that the disk is tilted at an angle of $\sim 22°$ with respect to the plane defined by the disk at large radii. Clearly such a configuration is unstable in an axisymmetric potential. We ask whether such a configuration might exist and be stable in a potential which is not axisymmetric, much in the same way that we have modeled phenomena such as the spindle galaxy (NGC 2685).

Several authors[9] have found a family of retrograde orbits which are tilted by the Coriolis force in a rotating triaxial potential. Since the disk in the galactic center is direct (sometimes called "prograde"), their orbits are not relevant to this particular problem.

We have looked at the existence and stability of 1-1-1 resonant elliptical orbits (all amplitudes nonzero and one oscillation out of phase with the other two) in a potential with harmonic plus quartic terms, i.e.,

$$\Phi = \frac{ax^2}{2} + \frac{by^2}{2} + \frac{fz^2}{2} - \frac{cx^2y^2}{2} - \frac{gx^2z^2}{2}$$
$$- \frac{hy^2z^2}{2} - \frac{dx^4}{4} - \frac{ey^4}{4} - \frac{Kz^4}{4} . \tag{1}$$

After finding a second integral of the motion (an invariant other than the energy), we can find the orbits and test for their stability using Liapunov's method.

The problem with the above prescription is that all the action takes place at an energy where the particles are barely bound (or even just unbound). It has however told us where to look and given some believable general results. For example, it predicts that the only stable one of these orbits is that which has the oscillation along the intermediate axis out of phase with the other two. Further, they do not exist in potentials like that used by Schwarzschild[5] even when different axial ratios are allowed. An example of a titled orbit is shown in Fig. 3. The values of the structure constants (given in the figure caption) should be considered arbitrary. We are currently thinking harder about performing more detailed model calculations to determine the central potential of the Galaxy.

Changing the orientations of disks has been a popular way of

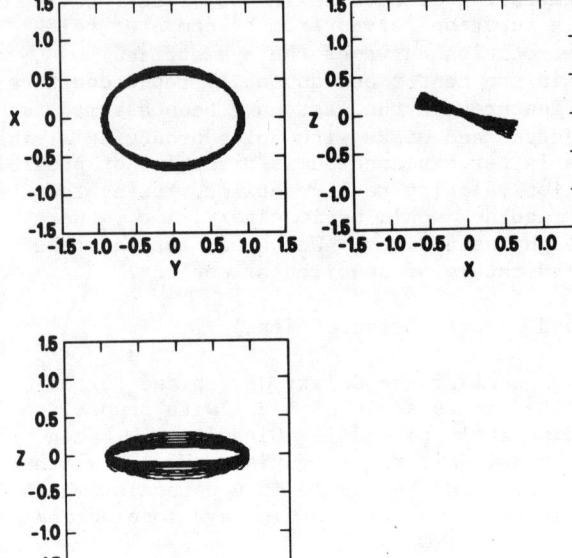

Fig. 3. Projection of the trajectory of a tilted orbit onto the three planes of symmetry of the potential. The structure constants (see Eq. (1)) are a = 0.24, b = 0.31, c = 0.16, d = 0.11, e = 0.21, f = 0.45, g = 0.48, h = 0.53, k = 0.85.

precessing jets. Kotanyi, et al.,[9] have proposed using the large-scale tumbling of a galaxy to do this. The problem is that this generally requires $V_{plasma} \sim$ velocity dispersion of the galaxy, resulting in excessive flow rates (in Cen A, $10^{11} M_\odot$ would have to be expelled). The idea might be salvaged by considering (the faster) tumbling of a more compact component.

c) Enhanced Fueling Rates in Active Galaxies

The main problem is supplying material to a nuclear source is shedding the angular momentum of the material. In a nonaxisymmetric potential, this is certainly less of a problem, as the angular momentum of an individual particle is not a conserved quantity.

We consider two basic alternatives for the fuel: stars and gas. In Schwarzschild's[5] model, over 80% of the stars are in box orbits. Each of these has a nonvanishing probability amplitude at the center. Clearly, the usual "loss-cone" problem is gone.[10] If the material is gas, the effect of any viscosity is amplified in the region where the eccentricity of the elliptical orbits is changing rapidly. The advantage of stars is that there are so many of them, though this is double-edged as there would seem to be no way to stem the flow to produce the observed luminosity evolution of QSOs.

REFERENCES

1. G. Illingworth, in NATO Proceedings of Cambridge Conference on Normal Galaxies, ed. T. N. Fall and D. Lynden-Bell, Cambridge University Press, Cambridge, 1981.
2. R. Leach, Ap. J., 248, 485, 1981.
3. G. Lake, in Photometry, Kinematics and Dynamics of Galaxies, ed. D. Evans, Univ. of Texas, Austin, 1979; G. Lake, 1981, in preparation.
4. G. Lake and C. Norman, 1981, in preparation.
5. M. Schwarzschile, Ap. J. 232, 236, 1979.
6. J. H. Lacy, F. Baas, C. H. Townes, and T. R. Geballe, Ap. J. (Letters) 227, L17, 1980.
7. J. Caldwell and J. P. Ostriker, 1981, preprint.
8. H. Liszt and W. B. Burton, Ap. J., 236, 779, 1980.
9. C. Kotanyi, T. Van Albada, and M. Schwarzschild, M.N.R.A.S., 1981, in press; J. Tohline and D. E. Osterbrock, 1981, preprint.
10. J. Silk and C. Norman, 1981, preprint.

## COMPARISON OF GALACTIC CENTER WITH OTHER GALAXIES

G. H. Rieke and M. J. Lebofsky
Steward Obs., Univ. of Ariz., Tucson, Az. 85721

### ABSTRACT

It is unlikely that the optical-infrared energetics of the Galactic Center are dominated by a single, nonthermal source as would be the case for a Seyfert or other kind of active galaxy. The energetics of this region can, however, be explained as the result of recent formation of massive stars; in this regard, our Galaxy may be analogous to M82, NGC 253, and similar galaxies.

### INTRODUCTION

The compact radio source in the Galactic Center and the various phenomena discovered from the gamma radiation demonstrate the importance of nonthermal processes in this region. However, the energy required to produce the far infrared emission and to ionize the gas is much greater than can be associated directly with either the compact radio source or the gamma ray source, leaving open the relation of the Galactic Center to the nuclei of Seyfert and other active galaxies. On a scale of a hundred parsecs or more, the energetics of the Galactic Center are dominated by young, luminous stars; viewed from a few megaparsecs, we would certainly find it difficult to find evidence for any other kind of activity. The uniquely high spatial resolution available to us in the Galactic Center, however, presents an opportunity to identify nonthermal activity on a relatively minor scale.

### NONTHERMAL MODELS FOR THE ENERGETICS OF THE GALACTIC CENTER

The presence of a super-compact, variable radio source coincident with the infrared source 16 has naturally suggested that source 16 is the optical-infrared counterpart of a compact, nonthermal source that provides much of the energy and ionizing flux within a few parsecs of the Galactic Center. This association can be tested in a number of ways. These comparisons are summarized in table 1 and described in more detail below.

The requirements on the source luminosity in table 1 are from the discussion of Becklin, Gatley, and Werner (1982) and those on the ionizing flux from Lacy et al. (1980). The source should have an apparent K (2.2 µm) magnitude no brighter than that of the unresolved core of source 16, which is K = 9.2; corrected for interstellar extinction of $A_v < 35$, this limit corresponds to K > 6.1. Since the Ne II emission (Lacy et al. 1980) and the hydrogen recombination emission (Neugebauer et al. 1978; Lebofsky, Rieke, and Thompson 1982) are found in virtually all directions from source 16 and at both positive and negative redshifts, it seems

Table 1.

|  | Nonthermal Source | Starburst |
|---|---|---|
| **Requirements** | | |
| Luminosity | > 0.5 to 1 X $10^7 L_\odot$ (from far IR obs.) | |
| | < 3 X $10^7 L_\odot$ (from grain temperature) | |
| Lyman cont. | > 1.5 X $10^{50}$ phot/sec | |
| Lyc. temp. | < 31000 K | < 35000 K |
| 2.2μm mag. | > 9.2 | |
| Others | source not beamed | |
| | color T map favors distributed lum. sources | |
| **Models and Performance** | radio loud QSO: Lyc. < $10^{45}$ | from presence of >7 red supgiants |
| | radio quiet QSO: Lyc. < 3 X $10^{47}$ | expect >80 blue supgiants; with |
| | optically thin synch. em. with monoenergetic electrons: Lyc. < 3 X $10^{48}$ | normal dist. over spec. type and lum. they give: |
| | optically thin viscous accretion disk: Lyc. < 3 X $10^{48}$ | Lum. > 1.5 X $10^7$ Lyc.> 1.9 X $10^{50}$ |
| | blackbody: Lyc. < 1 X $10^{50}$ | Lyc. T~35000 K Distributed lum. sources |

unlikely that the nonthermal source beams its ionizing flux and luminosity away from us so we do not detect it directly. Finally, we note that the width of the emission lines integrated across the Galactic Center is substantially less than the broad lines that would be expected of a classical Seyfert galaxy.

Since the compact radio source spectrum rises with increasing frequency, it is logical to compare it with flat spectrum extragalactic radio sources, most of which are also very compact and many of which are variable. The ratio of radio to infrared brightness of these sources is summarized in figure 1, from our unpublished measurements. A few galaxies where the infrared flux is known to be supplemented by strong thermal emission by extended clouds of dust are shown by dotted lines; these are not directly comparable with source 16. The ratio shown in the figure is not a strong function of source luminosity between 3 X $10^9$ and 3 X $10^{14}$ $L_\odot$, nor is it a strong function of the identification of the source with a galactic nucleus or stellar image, so long as appropriate corrections for redshift are made. Source 16 is at least 100 times too bright in the infrared to be associated with a similar object. Assuming a typical optical-infrared nonthermal continuum, going roughly as $\nu^{-1}$ with a high-frequency cutoff in accordance with the upper limit on the effective temperature, the luminosity and ionizing continuum from a "typical" nonthermal counterpart to the

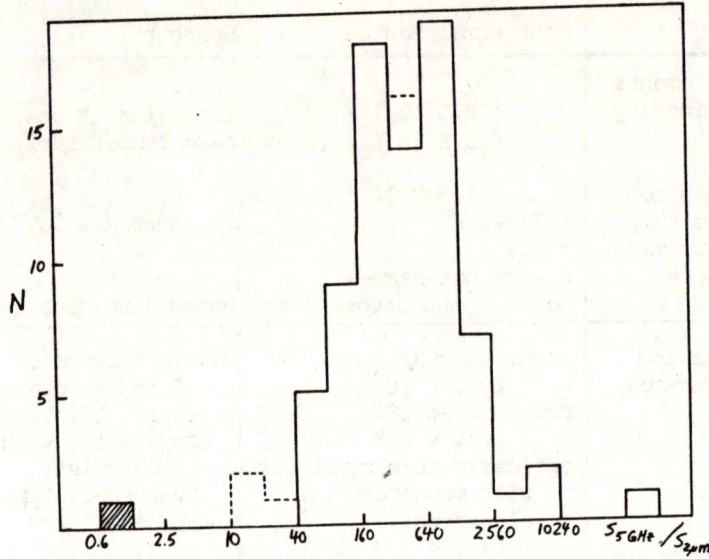

Fig. 1. Ratio of 5 GHz flux to 2.2 μm flux (frequency units) for extragalactic flat spectrum radio sources. Dashed lines show cases where the infrared flux is contaminated by extended thermal emission. Source 16 in the Galactic Center is indicated by the shaded box.

compact radio source can be estimated as shown in table 1. Both quantities are inadequate by orders of magnitude to dominate the processes in the Galactic Center.

The above arguments suggest that source 16 would have to be considered an extreme example of a radio quiet source, if it is to be compared with extragalactic nonthermal sources at all. Even in this case, it is not likely that the source provides the bulk of the luminosity and ionizing radiation in the region. For example, if the source has a spectrum falling as $\nu^{-1}$ to the appropriate high-frequency cutoff and normalized to the upper limit to the brightness at 2.2 μm, the predicted Lyman continuum is too weak by nearly three orders of magnitude.

Finally, a number of theoretically predicted spectra might be considered for source 16, although there are few if any extragalactic sources known with analogous behavior. The most favorable optically thin synchrotron source for meeting the requirements in table 1 would have monoenergetic electrons at constant pitch angle in a uniform magnetic field. In this case, the spectrum rises as $\nu^{1/3}$ at low frequencies and cuts off exponentially at high ones (see, e.g., Ginzburg and Syrovatskii

1965). Such a spectrum normalized to the upper limit on the brightness of the source falls short of the required Lyman continuum by nearly two orders of magnitude. Any synchrotron source powerful enough to dominate the energetics of the Galactic Center and not beamed away from us must be self-absorbed at wavelengths short of about 0.4 μm, a possibility that is very unattractive from a theoretical point of view. Another possibility is a thin, viscous accretion disk. However, this spectrum will also rise as $\nu^{1/3}$ and then cut off exponentially (Lynden-Bell and Pringle 1974), so the predicted Lyman continuum will be very similar to that for a monoenergetic synchrotron source. Even a blackbody spectrum rises too slowly from the brightness limit at 2.2 μm to provide adequate Lyman continuum emission within the constraint set by the upper limit on the effective temperature.

From these arguments, it seems quite unlikely that a single nonthermal source dominates the energetics of the inner few parsecs of the Galaxy.

## STARBURST MODELS FOR THE ENERGETICS OF THE GALACTIC CENTER

Since analogies with external galaxies whose nuclei are dominated by powerful nonthermal sources are not very successful for the Galactic Center, we will consider whether virtually all of the optical-IR phenomena in this region could result from stellar processes. First, we will compare the Galactic Center with other nearby galaxies whose energetics appear to be dominated by stars; we will then take a more detailed look at the evidence for recent rapid star formation in the central parsec of the Galaxy.

Figure 2 compares the 10 μm luminosity of the Galactic Center with that of the nuclei of other galaxies within about 10 Mpc. Although 10 μm is too short a wavelength to give an unbiased sample of the infrared luminosities of these galaxies, it is the longest wavelength where enough measurements are available to provide a reasonably complete sample. A difficulty in making the kind of comparison in figure 2 is that the observations with similar beam sizes refer to widely varying physical scales in the galaxies. Therefore, to construct figure 2 we first determined the dependence of 10 μm flux from the Galactic Center on aperture by correcting the measurements in Low et al. 1969, Rieke, Telesco, and Harper (1978), and Price and Walker (1976) for interstellar extinction assuming $A_v$ = 30. This relation was extended to larger apertures by means of the observed aperture dependence in the far infrared (e.g. Low and Aumann 1970). For each galaxy in figure 2 we then determined the physical region pertinent to the available 10 μm observation and took the ratio of the observed 10 μm flux to that from a physical region of the same size centered on the Galactic Center. Where the external galaxy was not detected, upper limits are indicated by downward pointing arrows. Extinction corrections were not made for any of the external galaxies, so in cases like M82 and NGC 253 the plotted ratios are lower limits.

Figure 2 makes it clear that the Galactic Center is quite

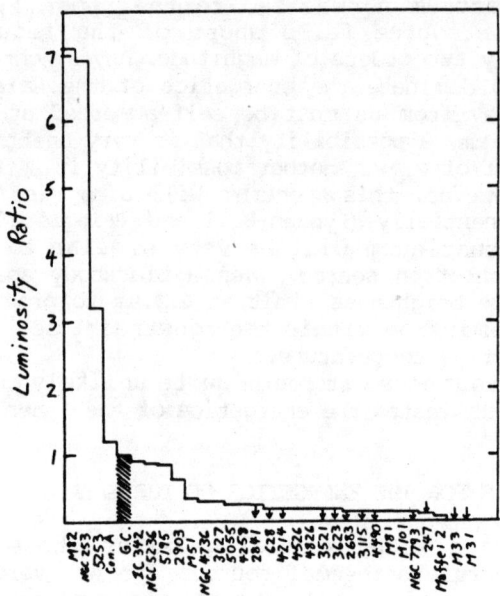

Fig. 2. Luminosity at 10 μm of large galaxies within 10 Mpc. Upper limits are indicated by downward pointing arrows; the Galactic Center is shaded.

energetic in the infrared when compared with the nuclei of a reasonably unbiased sample of large galaxies. Existing techniques would detect its 10 μm excess from a distance up to at least 30 Mpc. On the other hand, there are a number of galaxies, notably M82 and NGC 253, which are even more energetic and which appear to derive their power nearly entirely from recent bursts of star formation (see, e.g., Rieke et al. 1980).

Table 1 summarizes possible models for the Galactic Center in which recently formed stars provide the bulk of the energy and ultraviolet flux. The requirements are similar to those for nonthermal source models except that the upper limit on the temperature can be relaxed slightly because the ionizing sources in this type of model are presumably distributed through the volume of space.

Within a 3 arcmin region centered on the Galactic Center, there are at least 7 red supergiants---sources 7, 11, 12, 19, 22, 23, and 24 (Neugebauer et al. 1976; Wollman, Smith, and Larson 1981; Lebofsky, Rieke, and Tokunaga 1982). Additional supergiants are likely to be discovered in this region as infrared spectra are obtained for more stars (a number of candidates with appropriate photometric colors are already known). Humphreys (1978) has studied the population of very luminous stars in OB associations; she finds that there are about 11 times as many blue supergiants as red ones. Therefore, it seems very likely that there are 80 or more blue supergiants within the central few parsecs of the Galaxy. If these stars have the same distribution over spectral type and luminosity as was found by Humphreys for OB associations, the luminosity, ionizing flux, and effective temperature of the ionizing flux can be estimated as in table 1. These estimates are an exceedingly close match to the requirements for this region; the only possible discrepancy is in the effective temperature. However, the effective temperature of this particular population of hot stars is probably overestimated because the increased metallicity in the Galactic Center stars will soften their ultraviolet spectra (Shields and Tinsley 1976; Aitken, Griffiths, and Jones 1976). In addition, the absence of dense molecular clouds and other indications of appropriate conditions for ongoing star formation probably means that the episode leading to the current activity in the Galactic Center stopped some time ago; if massive stars have not formed during the past $10^6$ years or so, the hottest supergiants will have evolved to later spectral types and the constraint on the effective temperature of the ionizing flux will be met.

In addition to the indirect arguments above, there are at least two observations that support directly the importance of hot, luminous stars in the energetics of the Galactic Center. The first is that the color temperature of the heated dust in the Galactic Center shows many maxima, suggestive of a number of heating sources spread through the area. The second is that the cores of a number of the middle infrared sources have spectra which strongly suggest the presence of hot stars.

Figure 3 shows a map of the color temperature of the heated dust in this region. The contours at 120K and lower temperatures are based on a comparison of the 10 μm map of Rieke, Telesco, and Harper (1978) and the 56 μm map of Harvey, Campbell, and Hoffmann (1976). The relevant sections of these two maps have very nearly the same angular resolution (20" and 17" respectively). The relative positions of the maps were determined from the work of Becklin, Gatley, and Werner (1982). The contours in figure 3 above 120K are from a comparison of the 5 and 10 μm maps of Rieke and Low (1973) at 5.5" resolution with those of Rieke, Telesco, and Harper (1978) at 1.5" resolution. In addition, a color temperature peak is shown at the position of source 8 because the photometry of Becklin et al. (1978) indicates that one should be there. An important aspect of the color temperatures determined at 5 and 10 μm is that peaks are found within the smaller beam (see also the discussion by Rieke, Lebofsky, Deshpande, and Kemp 1982); since the typical sizes of the gas clouds are 5 to 8" (Lacy et al. 1980), this behavior requires that the temperature have a maximum within the cloud. Most models which heat the dust by Lyman alpha heating within the gas clouds or by some other externally driven mechanism will have difficulty producing this result.

The wavelength dependence of polarization for sources 1, 5, and 10 permits their spectra to be separated into a component

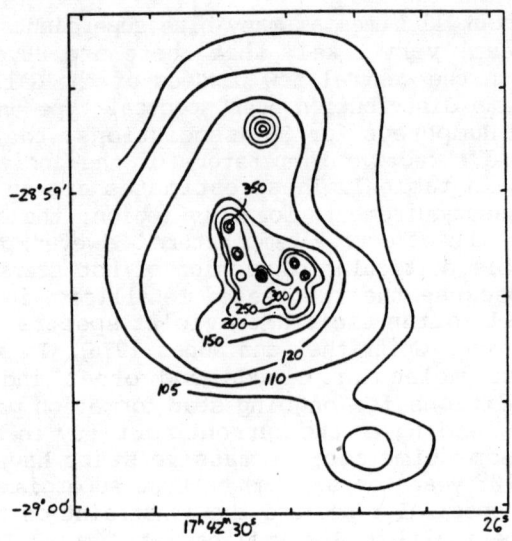

Fig. 3. Color temperature map of the Galactic Center. The contours below 150 K are at 20" resolution and from a comparison of 10 and 56 μm. The other contours are at 1.5" resolution (except for source 8) and from a comparison of 5 and 10 μm.

generated by thermal emission by aligned dust grains and an additional component that is dominant in the near infrared (Rieke, Lebofsky, Deshpande, and Kemp 1982). In all three cases, the spectrum of this additional component agrees closely with that expected from a star or stars reddened by $A_v$ = 30. Spectroscopy of the near infrared components of sources 1 and 13 shows both to have the shape between 1.9 and 2.4 μm expected of hot stars reddened by $A_v$ = 30 (Lebofsky, Rieke, and Thompson 1982); in addition, these spectra show no evidence for CO absorption bands in these stars, requiring that they be earlier than K in spectral type. It should be noted that the luminosities of these objects in the cores of the middle infrared sources are rather large for them to be single stars; instead, they are probably compact clusters or associations.

If the Galactic Center were the nucleus of a nearby external galaxy, in what ways could we study its nuclear starburst? Figure 2 demonstrates that it would be easily detectable by its infrared emission; the thermal radio flux and optical emission lines would be even easier to study. Since the extinction in the nucleus is relatively small (Becklin, Gatley, and Werner 1982), optical spectra would probably also show a filling in of the H and K break by the emission of hot stars. If the extinction were strong, as for M82 and NGC 253, we would have to make any direct detections of the luminous stellar population in the infrared. The possibilities for doing so are summarized in figure 4. In this figure, the CO band strengths for a quiescent stellar population are estimated by combining near infrared spectra of M81 and NGC 4736. A flux-weighted average supergiant spectrum has been determined for the Galactic Center and added to the spectrum of the quiescent population in the ratio observed between the supergiants and the extended emission in a 3 arcmin diameter region. The change in the overall spectrum would be very difficult to detect with existing techniques. However, if the supergiant population were increased in strength by a factor of ten, to correspond to the luminosity of the starbursts in M82 and NGC 253, figure 4 shows that a noticeable strengthening of the stellar CO bands should occur. Qualitatively, the resulting spectrum would be very similar to that actually observed from M82; the slightly different level of saturation of the bands can probably be explained by the bias of the known sample of Galactic Center supergiants toward relatively cool spectral types, since these stars have the largest bolometric corrections at 2 μm. This comparison with M82 strengthens the argument that the energetics of the Galactic Center are provided by rapid star

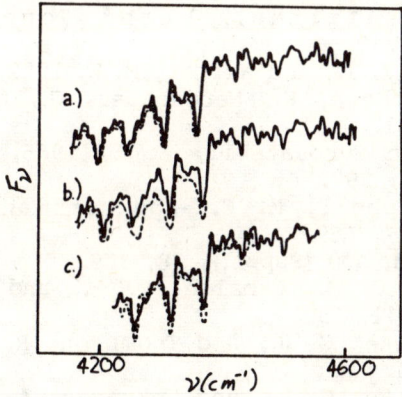

**Fig. 4.** Comparison of CO bands from cool stellar population in Galactic Center with those in other galaxies. a.) solid line: probable spectrum of quiescent stellar population; dashed line: expected spectrum of Galactic Center. b.) solid line: same as in a.; dashed line: spectrum with 10 times more flux from supergiants (spectra offset for clarity). c.) solid line: same as in a.; dashed line: spectrum of M82.

formation on a scale that is considerably larger than in the typical galactic nucleus, but which is still much weaker than the most extreme starburst galaxies known.

## SUMMARY

Although radio studies definitely establish the existence of a compact, nonthermal source in the Galactic Center, the phenomena which dominate this region in the optical and infrared do not appear to arise from this source. Therefore, our galaxy is not comparable with Seyfert galaxies or other galaxies which show nonthermal activity in the optical. The energetics of the inner few parsecs of the Galaxy can be explained virtually entirely by stellar processes, so long as very rapid star formation has occurred in this region in the recent past (up to $\sim 10^6$ years ago). Such a model gives a consistent explanation of: 1.) the energetics; 2.) the ionizing radiation field; 3.) the observed population of red supergiants; 4.) the thermal structure of the dust clouds emitting prominently near 10 µm; and 5.) the presence of relatively hot stars in the cores of these dust clouds. The Galactic Center is therefore similar to the starburst-powered nuclei of many spiral galaxies; it is in fact a relatively energetic example of such processes.

This work was supported by the National Science Foundation.

## REFERENCES

Aitken, D. K.,Griffiths, J.,and Jones, B. 1976, M.N.R.A.S.,176, 73P.
Becklin, E. E., Gatley, I., and Werner, M. W. 1982, ApJ., in press.
Becklin, E. E., Matthews, K., Neugebauer, G., and Willner, S. P. 1978, 219, 121.
Ginzburg, V. L., and Syrovatskii, S. I. 1965, Ann. Rev. Ast. and Astrophys.,3, 297.
Harvey, P. M., Campbell, M. F., and Hoffmann, W. F. 1976, ApJ(Letters), 205, L69.
Humphreys, R. M. 1978, ApJ (Suppl.), 38, 309.
Lacy, J. H., Towns, C. H., Geballe, T. R., and Hollenbach, D. J. 1980, ApJ., 241, 132.
Lebofsky, M. J., Rieke, G. H., and Thompson, R. I. 1982, to be submitted to ApJ.
Lebofsky, M. J., Rieke, G. H., and Tokunaga, A. T. 1982, to be submitted to ApJ.
Low, F. J., and Aumann, H. H. 1970, ApJ(Letters), 162, L79.
Low, F. J., Kleinmann, D. E., Forbes, F. F., and Aumann, H. H. 1969, ApJ(Letters), 157, L97.
Lynden-Bell, D.,and Pringle, J. E. 1974, M.N.R.A.S., 168, 603.
Neugebauer, G., Becklin, E. E., Beckwith, S., Matthews, K., and Wynn-Williams, C. G. 1976, ApJ(Letters), 205, L139.
Neugebauer, G., Becklin, E. E., Matthews, K., and Wynn-Williams, C. G. 1978, ApJ., 220, 149.

Price, S. D., and Walker, R. G. 1976, AFGL-TR-76-0208.
Rieke, G. H., Lebofsky, M. J., Deshpande, M., and Kemp, J. C. 1982, submitted to ApJ.
Rieke, G. H., Lebofsky, M. J., Thompson, R. I., Low, F. J., and Tokunaga, A. T. 1980, ApJ., 238, 24.
Rieke, G. H., and Low, F. J. 1973, ApJ., 184, 415.
Rieke, G. H., Telesco, C. M., and Harper, D. A. 1978, ApJ., 220, 149.
Shields, G. A., and Tinsley, B. M. 1976, ApJ., 203, 66.
Wollman, E. R., Smith, H. A., and Larson, H. P. 1981, ApJ., in press.

# RADIO OBSERVATIONS OF THE GALACTIC CENTER: FUTURE DIRECTIONS

Robert L. Brown
National Radio Astronomy Observatory*, Charlottesville, VA 22901

## ABSTRACT

The questions about the nature of the Galactic center region that one wishes to ask, and answer by means of future observations at radio frequencies, differ somewhat at different Galactocentric radii $R_g$. For the purposes of this discussion there appear to be four differentiable yet complementary regimes that bear consideration:

(1) $0.2 \leqslant R_g \leqslant 2$ kpc

(2) $3.0 \leqslant R_g \leqslant 200$ pc

(3) $0.3 \leqslant R_g \leqslant 3$ pc

(4) $R_g < 0.3$ pc

Here we briefly discuss each of these separately.

(1) $0.2 \leqslant R_g \leqslant 2$ kpc

The inner 10° of the Galaxy is a region rich in atomic and molecular material, the distribution and kinematics of which have been studied intensely by means of the λ21 cm line of HI for more than 20 years and in the λ2.6 mm 1-0 line of CO for nearly a decade. Lately the discussion has focused on whether the large-scale gas dynamics in the inner Galaxy reflects an episodic history of explosive events at the Galactic center that have been accompanied by violent gas ejection, the remains of which we now identify as specific kinematic features--the "3-kpc arm", the "+135 km s$^{-1}$ expanding arm"--or, alternatively, whether such features can be accounted for by a single, stable, kinematic structure such as a bar. To distinguish between these two possibilities requires extensive mapping over a large range in longitude, latitude and velocity. Here the most discriminating observations are those that help us define the velocity gradient within a particular kinematic feature; for this purpose it is necessary to employ multiple lines, or multiple molecular species, that allow one to sample the same line of sight with spectral lines of differing opacity.

*National Radio Astronomy Observatory is operated by Associated Universities, Inc., under contract with the National Science Foundation.

Because the area of sky to be studied is so large, ~100 square degrees, these investigations are best suited to single-dish spectroscopy. And, owing to the large fractional molecular abundance of the gas in the inner Galaxy, we anticipate that millimeter wavelength molecular spectroscopic observations will be particularly illuminating, illuminating not only the gas dynamical structures but also the chemistry and isotopic composition of the interstellar medium in the inner Galaxy. Fortunately several new millimeter-wave telescopes have recently been completed, or will be soon, that are ideal for spectroscopic investigations of the inner Galaxy; principal among them are the following:

(1)  Bell Laboratories 7m telescope,

(2)  University of Massachusetts 14.5m telescope,

(3)  IRAM 30m telescope,

(4)  Japanese (Nobeyama) 45m telescope.

Observational studies of the inner Galaxy should be an extremely active field of research in the next decade: several years from now the greatest need may be for sufficiently detailed gas dynamical models with which to compare the (anticipated) wealth of observational data.

(2)  $3 \leq R_g \leq 200$ pc

The molecular clouds closest to the Galactic center are truly enigmatic. The group at the Max-Planck-Institute for Radio Astronomy at Bonn--Güsten, Downes and their collaborators--have been particularly active in studying (principally) the centimeter-wave molecular lines from this region. They find the molecular clouds to have extremely large opacities, $A_v > 100$ magnitudes, uncommonly large velocity dispersions, $\Delta v \sim 25$ km s$^{-1}$, and, from ammonia observations, very high kinetic temperatures, $T \sim 100$ K. It is a challenge to future observations, and theoretical interpretations, to reconcile these observational data with the apparent long-lived stability of the clouds, the 2-10 μm observations toward Sgr A that suggest $A_v \sim 30$ magnitudes, and the mid and far-infrared observations from which a dust temperature much lower than the apparent gas temperature is inferred.

The dynamics of the ensemble of molecular clouds within 200 pc of the Galactic center is also puzzling. If they are indeed distributed within an (incomplete) expanding and rotating "molecular ring" then what event or process created and/or maintains this structure? Alternatively, if the ring is a kinematic, not spatial, structure--for example, part of the larger central bar-like gas flow--then why are the asymmetries so pronounced? The present body of observational data is insufficient to properly address these questions and, again, the theoretical modelling is wholly lacking.

(3) $0.3 \leq R_g \leq 3$ pc

Well interior to the molecular ring tidal forces are large and the gas motion is so turbulent that "clouds" cannot be long-lived. For this reason, together with the fact that the radiation field and cosmic-ray density appear enhanced in the inner parsec or so of the Galaxy, we infer that the gas in this region is distributed and it is largely ionized. Thus, to probe the inner Galaxy at radio frequencies requires that we make use of high resolution, interferometric, observations of the thermal radio continuum radiation and the radio recombination line radiation. Such studies are well under way employing both the Westerbork Synthesis Radio Telescope and the Very Large Array.

High resolution radio continuum observations are particularly useful in defining the density, and density distribution, of the thermal gas at the Galactic center unencumbered by line of sight extinction which plagues similar observations at visible and infrared wavelengths. The first results from the VLA reveal that the hot, $\sim 10^4$ K, thermal gas in the inner parsec or two of the Galaxy is not randomly distributed. Rather it appears to be organized into long, 10"-30", thin, 5", "ribbons" which define a spiral-like pattern. The ramifications of this result in terms of anisotropic mass ejection or response of the gaseous medium to the local gravitational potential warrant considerable further investigation.

The synthesis arrays have just begun to study the dynamics of the thermal material at the Galactic center by means of its radio recombination line radiation. These are challenging observations because one is attempting to study broad, $\Delta v \sim 500$ km s$^{-1}$, lines of low surface brightness. Nevertheless, such observations hold great promise both as a means for probing the thermodynamic structure of the thermal gas at the Galactic center and as a means for providing complementary information on the dynamical structure that has been suggested by the Ne II infrared observations.

(4) $R_g \leq 0.3$ pc

The nature of the object located in the inner 5" of the Galaxy is perhaps the final puzzle. Or, more likely, it is two puzzles. First, both the radio and the 10 μm infrared maps of the inner parsec of the Galaxy exhibit a symmetric brightness, but in each case the center of symmetry is not coincident with the non-thermal radio point source Sgr A*. Rather, the center of symmetry appears to lie ~3" south of Sgr A*. This immediately begs the question, is Sgr A* at the dynamical center of the Galaxy or is it simply a companion to a massive, dark, central object? In principle the measurements of the secular parallax of Sgr A* being

made by D. Backer and R. Sramek are capable of deciding this
issue--if Sgr A* exhibits no peculiar motion relative to the
extragalactic sky then, but only then, can it be regarded as
defining the dynamical center of the Galaxy--but the time-base of
the observations will need to be much longer before such a
judgment is forthcoming.

Secondly, the nature of the compact source Sgr A* itself
remains a mystery. While there appears to be no doubt that the
apparent angular size of the source varies with wavelength,
perhaps even as rapidly as $\theta \propto \lambda^2$, the interpretation of this
characteristic in terms of interstellar scattering or intrinsic
source structure remains ambiguous. Here the puzzle is compounded
by the fact that Sgr A* is time-variable. Repeated VLBI
observations, closely spaced in time at multiple frequencies, are
necessary to resolve this mystery in a satisfactory way. A
dedicated VLBI array, such as has been much discussed recently, is
the ideal, and perhaps necessary, instrument for these definitive
observations.

# FUTURE DIRECTIONS IN X-RAY/GAMMA-RAY OBSERVATIONS

D.A. Kniffen
NASA/Goddard Space Flight Center
Greenbelt, MD  20771

## ABSTRACT

X- and γ-ray observations from space platforms and ground observations over the last two decades have now extended observational astronomy to the shortest wavelengths of the electromagnetic spectrum. The penetrating nature of X-rays and γ-rays has opened up the potential for the direct observation of the highest energy processes in the center of our galaxy. To date, galactic center observations have been primarily exploratory, but the stage is now set for a major step forward in our knowledge of the galactic center region. The major orbiting satellites which will contribute to this knowledge are the Soviet-French Gamma-1 high energy γ-ray experiment scheduled for launch in 1983, ESA's EXOSAT, Japan's ASTRO-B, and West Germany's ROSAT, X-ray satellites scheduled for launchs in 1982, 1983, and 1986, respectively, and NASA's Gamma-Ray Observatory, currently planned for a 1988 launch. The latter includes four experiments which span the spectral range from .1 to $3 \times 10^4$ MeV with an order of magnitude improvement in sensitivity over any previous comparable observations. It is also expected that a vigorous balloon program will provide new techniques and observations which will strongly complement the satellite programs.

## INTRODUCTION

It is ironic that we have so little knowledge of some of the more basic characteristics of our own galaxy. Is ours a spiral galaxy? What is the constituency of the interstellar matter? Is there a black hole at the galactic center? Are the cosmic rays galactic in origin and is the cosmic ray intensity constant throughout the galaxy?

Upon reflection it is not so surprising that our knowledge is limited, since we reside in the galactic "suburbs" and must view the center through a great abundance of intervening gas and dust. Fortunately, with the advent of modern astronomy it is now possible to view the galactic center through the radio, infrared and X-ray to γ-ray windows. As we have heard at this conference this has already provided us with a new wealth of data with which to begin to unravel the mystery of our own galaxy. With X-rays and γ-rays we are already learning in the most direct way possible of the greatest energy transfers taking place in the most obscured regions of our galaxy.

In the X-ray range, the advent of satellite observations has provided a quantum leap in our knowledge of galactic sources in general, but more recently extended Einstein observations of the galactic center[1] have provided us with the best pictures yet of x-ray sources and extended emission from the galactic center. Balloon observations of high energy X-rays have also played a very important role.

In the γ-ray range, low energy γ-ray data from HEAO-2 and spectroscopic observations with balloons and HEAO-3 together with high energy data of SAS-2 and COS-B have opened the observations to the highest energy portions of the spectra. However, we have just begun the era of exploration in this wavelength range, and the observations of the 1980's should provide a wealth of new data which will greatly expand our knowledge of galactic emissions over the full range of the electromagnetic spectrum.

Approved satellite programs for the 1980's include EXOSAT, a European X-ray astronomy satellite scheduled for launch in 1982, Gamma-1, a Soviet-French high energy γ-ray experiment planned for launch in 1983, ROSAT, a German X-ray satellite with a planned 1986 launch and NASA's Gamma Ray Observatory planned for a 1988 launch. Combined with a vigorous balloon program and the development of new detector techniques, the prospects for a strong X-ray and γ-ray program for the 1980's looks reasonably good, given the constraints of ever tightening budgets.

The following sections will review the capabilities which will be available for X- and γ-ray observations in the 1980's.

## X-RAYS

Budgetary contraints in NASA's high energy astrophysics program have delayed the approval of any major new X-ray astronomy programs, including Explorers. The major candidates for the 1980's include the Advanced X-ray Astrophysics Facility (AXAF), a large high-resolution grazing incidence X-ray telescope, and the X-ray Timing Explorer (XTE), a mission to study the temporal variations of X-ray sources. Both of these missions will open new realms of X-ray astronomy to study. AXAF, a major national facility with a 10-15 year planned lifetime will provide vastly improved sensitivity and spatial resolution to that of the very successful Einstein Observatory (HEAO-2). Proposals received in response to NASA's Announcement of Opportunity for XTE are currently being evaluated so the payload is not yet known, but the emphasis will be on high sensitivity observations of short and long term variability of X-ray sources. The two missions have been discussed in detail at the December 13, 1980 Uhuru Memorial Symposium[2].

For the near term the most active X-ray astronomy program is in Europe where the major programs are EXOSAT and ROSAT. EXOSAT will be placed in a highly eccentric orbit ($2 \times 10^5$ by 500 km), allowing both a lunar occulation mode, which provides source positions in the 1-10 arc second range, and an offset pointing mode with spectral coverage from ~0.04 to 60 keV with three major instruments. The major improvements of this mission are for observations of galactic and extragalactic X-ray sources brighter than a few tenths of a mCrab and for the imaging of objects with soft X-ray spectra.

ROSAT (Röntgen Satellite) consists of a large X-ray mirror system with imaging proportional counters in the focal plane. The main objective of this mission is to provide the first all sky survey with an imaging telescope with over a thousand times the sensitivity of Uhuru.

Scheduled for a 1983 launch is the Japanese ASTRO-B, which consists of an array of gas scintillation counters to study the time variability of X-ray sources with improved energy resolution over conventional proportional counter observations.

The overall characteristics of these missions are summarized in Table I. The Proceedings of the Uhuru Memorial Symposium[2] contain detailed discussions of each of them.

Table I
X-Ray Satellite for the 1980's

| Program | Approx. Launch Date | Energy Range | Approximate Posit. Resolution | Sensitivity |
|---|---|---|---|---|
| EXOSAT | Mid 1982 | 0.04 to 60 KeV | few arc sec | ~ Einstein |
| ASTRO-B | Early 1983 | 1 to 60 KeV | moderate | moderate |
| ROSAT | 1986 | 0.5 to 3 KeV | 5 arc sec | $\gtrsim$ Einstein |
| AXAF | unknown | 0.1 to 8 keV | < 0.5 arc sec | ~ Einstein/$10^2$ |

The remainder of the planned orbital program involves Spacelab payloads. Several X- and γ-ray experiments have been selected for definition for possible inclusion on Spacelab missions. The most likely candidates for the near term are the three X-ray experiments selected for definition study for OSS-2. These are LAMAR, DXS, and BBXRT. DXS is a soft X-ray Bragg Spectrometer for studying the X-ray background in the hot interstellar medium. Dr. W. Kraushaar of the University of Wisconsin is the Principal Investigator. BBXRT, a Broadband X-ray Telescope with Dr. Peter Serlemitsos of the Goddard Space Flight Center as Principal Investigator, is a telescope designed primarily to study the spectra of quasars in the 0.5 to 10 keV band. The most pertinent to galactic center studies is the Large Area Modular Array of Reflectors (LAMAR) which is a sky survey instrument covering the 0.1 to 6 keV range. It will survey with good angular resolution a large area of the sky to a sensitivity 100 times better than Uhuru.

The orbital programs outlined above leave a spectral gap between about 60 keV and the approximately 100 keV threshold of the Gamma-Ray Observatory. The only opportunities for this energy range appear to be with the use of scientific balloons. The 20-

200 keV range observable with this technique includes the expected energies for the cyclotron features resulting from the quantum mechanical quantization of electron energy levels in the intense magnetic fields surrounding neutron stars. In addition, measurements will be made to observe very fast variability from some galactic objects.

## γ-RAYS

Only two orbiting γ-ray missions appear likely for the 1980's. The first, scheduled for launch in 1983, is the French-Soviet Gamma-1 spark chamber high energy γ-ray telescope. The telescope consists of wide gap spark chambers with a time-of-flight/cerenkov trigger telescope. The instrument is optimized to give good angular resolution for discrete sources. For this reason a passive coded aperture is included which can be inserted or removed on command. A circular (∼ 250 km) low-Earth orbit (inclination ≲ 65°) is planned. The characteristics of this instrument are given in Table II.

### Table II
### Gamma-I Characteristics

| | |
|---|---|
| Energy Range | > 50 MeV |
| Sensitive Area | 1400 $cm^2$ |
| Geometric Factor | 120 $cm^2$ ster |
| Detection Efficiency > 150 MeV | 20 percent |
| Angular Resolution at 100 MeV | ∼ 2 degrees |
| with Coded Aperture | 1-5 arc minutes |

NASA's major thrust in γ-ray astronomy in the 1980's is the Gamma Ray Observatory (GRO). This approved shuttle launched mission is currently scheduled for launch in 1988 with a complement of four instruments designed to cover the energy range from about 60 keV to $3 \times 10^4$ MeV. Figure 1 is an artists conception of GRO in orbit. The sensitivity over the full range of energies is over an order of magnitude better than those of previous γ-ray missions.

Equally as important is the vastly improved positional resolution for discrete objects. Since no single instrument can cover the full range of objectives, each of the four is designed to emphasize a specific aspect of the mission objectives.

The Orienting Scintillation Spectrometer Experiment (100 keV to 10 MeV): This experiment utilizes four large actively-shielded and passively-collimated Sodium Iodide (NaI) Scintillation detectors, with a 5°x11° FWHM field-of-view. The large area detectors provide excellent sensitivity for both γ-ray line and continuum emissions. An offset pointing system modulates the celestial source contributions to allow background substraction. It also permits observations of off-axis sources such as transient

Figure 1: The Gamma Ray Observatory
(Photo Courtesy of TRW)

phenomena and solar flares without impacting the planned Observatory viewing program. Dr. J. Kurfess of the Naval Research Laboratory is the Principal Investigator.

The Imaging Compton Telescope: This instrument is based on a newly established concept of γ-ray detection in the 1-30 MeV range. It employs the unique signature of a two-step absorption of the γ-ray, i.e., a Compton collision in the first detector followed by total absorption in a second detector element. This method, in combination with effective charged particle shield detectors, results in a more efficient suppression of the otherwise inherent instrumental background.

Spatial resolution in the two detectors together with the well defined geometry of the Compton interaction permits the reconstruction of the sky image over a wide field-of-view (~ 1 steradian) with a resolution of a few degrees. In addition, the instrument has the capability of searching for polarization of the radiation. The instrument has good capabilities for the search for weak sources, weak galactic features and for the search for spectral and spatial features in the extragalactic diffuse radiation. Dr. V. Schönfelder of the Max-Planck Institut (MPE), Garching, is Principal Investigator of this investigation.

The High Energy Gamma Ray Telescope: The High Energy Gamma-ray Telescope is designed to cover the energy range from 20 MeV to $30 \times 10^3$ MeV. The instrument uses a multi-thin-plate spark chamber to detect γ-rays by the electron-positron pair process. A total energy counter using a NaI(Tℓ) shower counter is placed beneath the instrument to provide good energy resolution over a wide dynamic range. The instrument is covered by a plastic scintillator anticoincidence dome to prevent readout on events not associated with γ-rays. The combination of high energies and good spatial resolution in this instrument provides the best source positions of any GRO instrument. Co-principal Investigators of this investigation are Drs. C. Fichtel of the Goddard Space Flight Center, R. Hofstadter of the Stanford University, and K. Pinkau of the Max-Planck-Institut.

The Burst and Transient Source Experiment: The Burst and Transient Source Experiment for the GRO is designed to continuously monitor a large fraction of the sky for a wide range of types of transient γ-ray events. The monitor consists of eight wide field detector modules. Four have the same viewing path as the other telescopes on GRO and four are on the bottom side of the instrument module viewing the opposite hemisphere. This arrangement provides maximum continuous exposure to the unobstructed sky. The capability provides for 0.1 msec time resolution, a burst location accuracy of about a degree and a sensitivity of $6 \times 10^{-8}$ erg/cm$^2$ for a 10 sec burst. Dr. J. Fishman of the Marshall Space Flight Center is Principal Investigator.

The salient features of the four experiments are summarized in Table III. Each instrument represents a significant step forward over its predecessors. For example, the sensitivity for line gamma-ray detection has been improved by more than an order of magnitude over the HEAO-A4 and HEAO-C1 instruments. The continuum sensitivity in the MeV range is typically improved by a factor of twenty or more. Improvements of about an order of magnitude in source location capability are also expected due to the improved instruments and the greatly increased exposure factors. The addition of a massive NaI calorimeter crystal has markedly improved the energy resolution (a factor of $\gtrsim 2$ better than SAS-2) in the > 100 MeV range and extended the range to 30 GeV. Also in this range the total effective area (i.e., area X geometry factor) is 25 times greater than that of COS-B.

The 12500 kg observatory will be launched aboard the Space Shuttle. It will be placed in a 400 km circular 28.5 degree inclination orbit where it will remain for a two year mission lifetime. Although three-axis stabilized, the primary objective of the first year will be a full sky survey.

In the decade of the 1980's significant contributions to γ-ray astronomy will be made from platforms flown on scientific balloons. A strong balloon program is crucial not just for the observational results which can be obtained but especially because of the opportunity for developing new detector techniques. As the capability is extended to provide long duration flights of days to weeks, ballooning will be a very important part of a complete program in γ-ray astronomy. Tables IV and V summarize the continuing observational and developmental programs.

The objective of the instrument developments is to provide improved source locations at all energies from hard X-rays to γ-rays in order to improve the opportunity for source identifications. A specific example important to galactic center observations is the coded aperture telescope developed at the University of New Hampshire[3]. In this instrument a coded mask sits in front of an array of Bismuth Germanate detectors which serve as the imaging detector. The combination covers the 100 keV to 5 MeV

Table III
SUMMARY OF GRO DETECTOR CHARACTERISTICS

| | OSSE | COMPTEL | EGRET | BATSE |
|---|---|---|---|---|
| Energy Range (MeV) | 0.10 to 10.0 | 1.0 to 30.0 | 20 to $3 \times 10^4$ | 0.05 to 0.60 |
| Energy Resolution | 8.0% at 0.66 MeV | 5 – 8% | 15% | 35% at 0.1 MeV |
| Maximum Effective Area ($cm^2$ efficiency) | 2310 | 50 | 2000 | 5500 |
| Position Resolution (strong source) | 10 arc min square error box (special mode) | 7.5 arc min (1σ radius) | 5 arc min (1σ radius) | 1° |
| Maximum Effective Geometric Factor ($cm^2 \cdot sr$ effeciency) | 12 | 30 | 1000 | 15000 |
| Estimated Threshold Line (source sensitivity) | $2 \times 10^{-5} cm^{-2} s^{-1}$ | $3 \times 10^{-5}$ to $3 \times 10^{-6}$ | | 0.1 Crab-transient |
| Continuum | $\sim 3 \times 10^{-5} cm^{-2} s^{-1}$ | $5 \times 10^{-5} cm^{-2} s^{-1}$ | $5 \times 10^{-8} cm^{-2} s^{-1}$ | $6 \times 10^{-8} erg\, cm^{-2}$-burst |
| Weight (Kg) | 1730 | 1477 | 1708 | 570 |

Table IV
Current Observational Balloon Programs

| Detector Type | Objectives |
|---|---|
| Imaging Compton Telescope | Observations in the relatively unexplored medium energy γ-ray region. |
| Scintillation Telescope (Hard X-ray/Low energy γ-ray) | Broad-line observations of sources including .511 MeV galactic center line, cyclotron resonance, etc. |
| Collimated Compton Telescope | Low background point source observations in the low to medium energy γ-ray region. |
| Germanium Spectrometer | High energy resolution narrow line spectroscopy. Especially .511 MeV galactic center line. |
| Spark Chambers | Observations of high energy γ-rays. |

Table V
Detectors Being Developed

| Technique | Objective |
|---|---|
| Coded Aperture (with all combinations of positional detectors) | Good Positional Resolution at all γ-ray energies |
| Spark Chamber | Good Source Locations at High Energies ($\gtrsim$ 150 MeV) |
| Finely Collimated Compton Telescope | Good Angular Resolution at Medium Energies (1 to 30 MeV) |

range with a source location accuracy of ±1°, a major improvement for this region of the spectrum. A 1982 balloon observation of the galactic center is planned.

## SUMMARY

It is clear that progress in X- and γ-ray astronomy depends upon continued opportunities with various types of space platforms. On the other hand, these observations are of very high priority because together with a continued strong observational program at other wavelengths, astronomers will have the opportunity to observe objects over the full range of the electromagnetic spectrum. It appears that such a program will occur for the first time in the 1980's, even though bugetary contraints will result in a slower than desired development of new capabilities. Nevertheless, unprecedented growth in our knowledge of our own galaxy and especially the galactic center should occur by the end of the decade. The search for time variations in .511 MeV line from the galactic center will continue. The detailed observations of galactic sources in general, especially in the confusing central region of the galaxy will lead to dramatic improvements in our knowledge of the energy spectra (including lines), the source locations and their time variability. The question of the variation of the cosmic rays over the galaxy and the relative contribution and nature of galactic sources will be addressed with continued high energy γ-ray observations.

## REFERENCES
1. M. G. R. Willingale, J.E. Grindlay and P. Hertz, Ap. J. 250, 142, (1981).
2. Journal of the Washington Academy of Sciences, 71, June 1981.
3. M. L. McConnell, D. J. Forrest, E. L. Chupp, and P. P. Dunphy, IEEE Nuclear Science Symposium, San Francisco, October 1981.

## AIP Conference Proceedings

|  |  | L.C. Number | ISBN |
|---|---|---|---|
| No.1 | Feedback and Dynamic Control of Plasmas | 70-141596 | 0-88318-100-2 |
| No.2 | Particles and Fields - 1971 (Rochester) | 71-184662 | 0-88318-101-0 |
| No.3 | Thermal Expansion - 1971 (Corning) | 72-76970 | 0-88318-102-9 |
| No.4 | Superconductivity in d-and f-Band Metals (Rochester, 1971) | 74-18879 | 0-88318-103-7 |
| No.5 | Magnetism and Magnetic Materials - 1971 (2 parts) (Chicago) | 59-2468 | 0-88318-104-5 |
| No.6 | Particle Physics (Irvine, 1971) | 72-81239 | 0-88318-105-3 |
| No.7 | Exploring the History of Nuclear Physics | 72-81883 | 0-88318-106-1 |
| No.8 | Experimental Meson Spectroscopy - 1972 | 72-88226 | 0-88318-107-X |
| No.9 | Cyclotrons - 1972 (Vancouver) | 72-92798 | 0-88318-108-8 |
| No.10 | Magnetism and Magnetic Materials - 1972 | 72-623469 | 0-88318-109-6 |
| No.11 | Transport Phenomena - 1973 (Brown University Conference) | 73-80682 | 0-88318-110-X |
| No.12 | Experiments on High Energy Particle Collisions - 1973 (Vanderbilt Conference) | 73-81705 | 0-88318-111-8 |
| No.13 | $\pi$-$\pi$ Scattering - 1973 (Tallahassee Conference) | 73-81704 | 0-88318-112-6 |
| No.14 | Particles and Fields - 1973 (APS/DPF Berkeley) | 73-91923 | 0-88318-113-4 |
| No.15 | High Energy Collisions - 1973 (Stony Brook) | 73-92324 | 0-88318-114-2 |
| No.16 | Causality and Physical Theories (Wayne State University, 1973) | 73-93420 | 0-88318-115-0 |
| No.17 | Thermal Expansion - 1973 (lake of the Ozarks) | 73-94415 | 0-88318-116-9 |
| No.18 | Magnetism and Magnetic Materials - 1973 (2 parts) (Boston) | 59-2468 | 0-88318-117-7 |
| No.19 | Physics and the Energy Problem - 1974 (APS Chicago) | 73-94416 | 0-88318-118-5 |
| No.20 | Tetrahedrally Bonded Amorphous Semiconductors (Yorktown Heights, 1974) | 74-80145 | 0-88318-119-3 |
| No.21 | Experimental Meson Spectroscopy - 1974 (Boston) | 74-82628 | 0-88318-120-7 |
| No.22 | Neutrinos - 1974 (Philadelphia) | 74-82413 | 0-88318-121-5 |
| No.23 | Particles and Fields - 1974 (APS/DPF Williamsburg) | 74-27575 | 0-88318-122-3 |
| No.24 | Magnetism and Magnetic Materials - 1974 (20th Annual Conference, San Francisco) | 75-2647 | 0-88318-123-1 |
| No.25 | Efficient Use of Energy (The APS Studies on the Technical Aspects of the More Efficient Use of Energy) | 75-18227 | 0-88318-124-X |

| | | | |
|---|---|---|---|
| No.26 | High-Energy Physics and Nuclear Structure<br>- 1975 (Santa Fe and Los Alamos) | 75-26411 | 0-88318-125-8 |
| No.27 | Topics in Statistical Mechanics and Biophysics:<br>A Memorial to Julius L. Jackson<br>(Wayne State University, 1975) | 75-36309 | 0-88318-126-6 |
| No.28 | Physics and Our World: A Symposium in Honor<br>of Victor F. Weisskopf (M.I.T., 1974) | 76-7207 | 0-88318-127-4 |
| No.29 | Magnetism and Magnetic Materials - 1975<br>(21st Annual Conference, Philadelphia) | 76-10931 | 0-88318-128-2 |
| No.30 | Particle Searches and Discoveries - 1976<br>(Vanderbilt Conference) | 76-19949 | 0-88318-129-0 |
| No.31 | Structure and Excitations of Amorphous Solids<br>(Williamsburg, VA., 1976) | 76-22279 | 0-88318-130-4 |
| No.32 | Materials Technology - 1976<br>(APS New York Meeting) | 76-27967 | 0-88318-131-2 |
| No.33 | Meson-Nuclear Physics - 1976<br>(Carnegie-Mellon Conference) | 76-26811 | 0-88318-132-0 |
| No.34 | Magnetism and Magnetic Materials - 1976<br>(Joint MMM-Intermag Conference, Pittsburgh) | 76-47106 | 0-88318-133-9 |
| No.35 | High Energy Physics with Polarized Beams and<br>Targets (Argonne, 1976) | 76-50181 | 0-88318-134-7 |
| No.36 | Momentum Wave Functions - 1976 (Indiana University) | 77-82145 | 0-88318-135-5 |
| No.37 | Weak Interaction Physics - 1977 (Indiana University) | 77-83344 | 0-88318-136-3 |
| No.38 | Workshop on New Directions in Mossbauer<br>Spectroscopy (Argonne, 1977) | 77-90635 | 0-88318-137-1 |
| No.39 | Physics Careers, Employment and Education<br>(Penn State, 1977) | 77-94053 | 0-88318-138-X |
| No.40 | Electrical Transport and Optical Properties of<br>Inhomogeneous Media (Ohio State University, 1977) | 78-54319 | 0-88318-139-8 |
| No.41 | Nucleon-Nucleon Interactions - 1977 (Vancouver) | 78-54249 | 0-88318-140-1 |
| No.42 | Higher Energy Polarized Proton Beams<br>(Ann Arbor, 1977) | 78-55682 | 0-88318-141-X |
| No.43 | Particles and Fields - 1977 (APS/DPF, Argonne) | 78-55683 | 0-88318-142-8 |
| No.44 | Future Trends in Superconductive Electronics<br>(Charlottesville, 1978) | 77-9240 | 0-88318-143-6 |
| No.45 | New Results in High Energy Physics - 1978<br>(Vanderbilt Conference) | 78-67196 | 0-88318-144-4 |
| No.46 | Topics in Nonlinear Dynamics (La Jolla Institute) | 78-057870 | 0-88318-145-2 |
| No.47 | Clustering Aspects of Nuclear Structure and<br>Nuclear Reactions (Winnepeg, 1978) | 78-64942 | 0-88318-146-0 |
| No.48 | Current Trends in the Theory of Fields<br>(Tallahassee, 1978) | 78-72948 | 0-88318-147-9 |
| No.49 | Cosmic Rays and Particle Physics - 1978<br>(Bartol Conference) | 79-50489 | 0-88318-148-7 |

## AIP Conference Proceedings

|  |  | | |
|---|---|---|---|
| No. 50 | Laser-Solid Interactions and Laser Processing - 1978 (Boston) | 79-51564 | 0-88318-149-5 |
| No. 51 | High Energy Physics with Polarized Beams and Polarized Targets (Argonne, 1978) | 79-64565 | 0-88318-150-9 |
| No. 52 | Long-Distance Neutrino Detection - 1978 (C.L. Cowan Memorial Symposium) | 79-52078 | 0-88318-151-7 |
| No. 53 | Modulated Structures - 1979 (Kailua Kona, Hawaii) | 79-53846 | 0-88318-152-5 |
| No. 54 | Meson-Nuclear Physics - 1979 (Houston) | 79-53978 | 0-88318-153-3 |
| No. 55 | Quantum Chromodynamics (La Jolla, 1978) | 79-54969 | 0-88318-154-1 |
| No. 56 | Particle Acceleration Mechanisms in Astrophysics (La Jolla, 1979) | 79-55844 | 0-88318-155-X |
| No. 57 | Nonlinear Dynamics and the Beam-Beam Interaction (Brookhaven, 1979) | 79-57341 | 0-88318-156-8 |
| No. 58 | Inhomogeneous Superconductors - 1979 (Berkeley Springs, W.V.) | 79-57620 | 0-88318-157-6 |
| No. 59 | Particles and Fields - 1979 (APS/DPF Montreal) | 80-66631 | 0-88318-158-4 |
| No. 60 | History of the ZGS (Argonne, 1979) | 80-67694 | 0-88318-159-2 |
| No. 61 | Aspects of the Kinetics and Dynamics of Surface Reactions (La Jolla Institute, 1979) | 80-68004 | 0-88318-160-6 |
| No. 62 | High Energy $e^+e^-$ Interactions (Vanderbilt, 1980) | 80-53377 | 0-88318-161-4 |
| No. 63 | Supernovae Spectra (La Jolla, 1980) | 80-70019 | 0-88318-162-2 |
| No. 64 | Laboratory EXAFS Facilities - 1980 (Univ. of Washington) | 80-70579 | 0-88318-163-0 |
| No. 65 | Optics in Four Dimensions - 1980 (ICO, Ensenada) | 80-70771 | 0-88318-164-9 |
| No. 66 | Physics in the Automotive Industry - 1980 (APS/AAPT Topical Conference) | 80-70987 | 0-88318-165-7 |
| No. 67 | Experimental Meson Spectroscopy - 1980 (Sixth International Conference, Brookhaven) | 80-71123 | 0-88318-166-5 |
| No. 68 | High Energy Physics - 1980 (XX International Conference, Madison) | 81-65032 | 0-88318-167-3 |
| No. 69 | Polarization Phenomena in Nuclear Physics - 1980 (Fifth International Symposium, Santa Fe) | 81-65107 | 0-88318-168-1 |
| No. 70 | Chemistry and Physics of Coal Utilization - 1980 (APS, Morgantown) | 81-65106 | 0-88318-169-X |
| No. 71 | Group Theory and its Applications in Physics - 1980 (Latin American School of Physics, Mexico City) | 81-66132 | 0-88318-170-3 |
| No. 72 | Weak Interactions as a Probe of Unification (Virginia Polytechnic Institute - 1980) | 81-67184 | 0-88318-171-1 |
| No. 73 | Tetrahedrally Bonded Amorphous Semiconductors (Carefree, Arizona, 1981) | 81-67419 | 0-88318-172-X |
| No. 74 | Perturbative Quantum Chromodynamics (Tallahassee, 1981) | 81-70372 | 0-88318-173-8 |

| | | | |
|---|---|---|---|
| No. 75 | Low Energy X-ray Diagnostics-1981 (Monterey) | 81-69841 | 0-88318-174-6 |
| No. 76 | Nonlinear Properties of Internal Waves (La Jolla Institute, 1981) | 81-71062 | 0-88318-175-4 |
| No. 77 | Gamma Ray Transients and Related Astrophysical Phenomena (La Jolla Institute, 1981) | 81-71543 | 0-88318-176-2 |
| No. 78 | Shock Waves in Condensed Matter - 1981 (Menlo Park) | 82-70014 | 0-88318-177-0 |
| No. 79 | Pion Production and Absorption in Nuclei - 1981 (Indiana University Cyclotron Facility) | 82-70678 | 0-88318-178-9 |
| No. 80 | Polarized Proton Ion Sources (Ann Arbor, 1981) | 82-71025 | 0-88318-179-7 |
| No. 81 | Particles and Fields - 1981: Testing the Standard Model (APS/DPF, Santa Cruz) | 82-71156 | 0-88318-180-0 |
| No. 82 | Interpretation of Climate and Photochemical Models, Ozone and Temperature Measurements (La Jolla Institute, 1981) | 82-071345 | 0-88318-181-9 |
| No. 83 | The Galactic Center (Cal. Inst. of Tech., 1982) | 82-071635 | 0-88318-182-7 |

RAYMOND H. FOGLER LIBRARY

DATE DUE

BOOKS ARE SUBJECT TO
RECALL AFTER TWO WEEKS

FEB 2 4 1984

Renewal: March 30, 1984